SpringerBriefs in Molecular Science

SpringerBriefs in Molecular Science present concise summaries of cutting-edge research and practical applications across a wide spectrum of fields centered around chemistry. Featuring compact volumes of 50 to 125 pages, the series covers a range of content from professional to academic. Typical topics might include:

- A timely report of state-of-the-art analytical techniques
- A bridge between new research results, as published in journal articles, and a contextual literature review
- A snapshot of a hot or emerging topic
- An in-depth case study
- A presentation of core concepts that students must understand in order to make independent contributions

Briefs allow authors to present their ideas and readers to absorb them with minimal time investment. Briefs will be published as part of Springer's eBook collection, with millions of users worldwide. In addition, Briefs will be available for individual print and electronic purchase. Briefs are characterized by fast, global electronic dissemination, standard publishing contracts, easy-to-use manuscript preparation and formatting guidelines, and expedited production schedules. Both solicited and unsolicited manuscripts are considered for publication in this series.

Ramachandra Naik · H. P. Nagaswarupa ·
B. H. M. Darukesha · D. M. Tejashwini

Advances in Space Radiation Detection

Novel Nanomaterials and Techniques

 Springer

Ramachandra Naik 🔘
Department of Applied Sciences
New Horizon College of Engineering
Bengaluru, Karnataka, India

B. H. M. Darukesha
Department of Space
Indian Space Research Organisation
Bengaluru, Karnataka, India

H. P. Nagaswarupa
Department of Studies in Chemistry
Davangere University
Davangere, Karnataka, India

D. M. Tejashwini
Department of Studies in Chemistry
Davangere University
Davangere, Karnataka, India

ISSN 2191-5407   ISSN 2191-5415 (electronic)
SpringerBriefs in Molecular Science
ISBN 978-3-031-74550-8   ISBN 978-3-031-74551-5 (eBook)
https://doi.org/10.1007/978-3-031-74551-5

This Springer imprint is published by the registered company Springer Nature Switzerland AG
The registered company address is: Gewerbestrasse 11, 6330 Cham, Switzerland

If disposing of this product, please recycle the paper.

Preface

Space exploration has always captivated human imagination, driving us to venture into the unknown realms of our universe. As we push the boundaries of space travel, understanding and mitigating the risks associated with space radiation becomes increasingly critical. Space radiation presents significant challenges to both spacecraft and astronauts, making the development of advanced radiation detection systems essential for future missions.

This book, *Advances in Space Radiation Detection*, is a comprehensive exploration of the current state and future directions in this vital field. It is designed to serve as a valuable resource for researchers, engineers, and students who are working toward enhancing our capabilities to detect, measure, and manage space radiation.

In Chap. 1, we lay the foundation by introducing the fundamental concepts and significance of space radiation detection. Understanding these basics is crucial for appreciating the complexity and importance of the subsequent advancements discussed throughout the book.

Chapter 2 delves into the fundamentals of nanomaterial synthesis, providing insight into the methods and techniques used to create the innovative materials that play a crucial role in modern radiation detection systems.

The characterization of metal oxide nanomaterials is explored in Chap. 3, where we examine the properties and behaviors of these materials under different conditions. This chapter highlights the importance of precise characterization in developing effective radiation detectors.

Chapter 4 focuses on the thermoluminescence properties of metal oxide nanomaterials. This section explains how these materials emit light when heated, a property that can be harnessed to improve radiation detection.

Chapter 5 explores the integration of zirconium dioxide (ZrO_2) nanomaterials into scintillation detectors. This chapter provides insights into how these materials enhance the performance of scintillation detectors, which are essential for detecting high-energy radiation.

In Chap. 6, we address methods to enhance the efficiency of plastic scintillators, which are widely used in radiation detection applications. Innovations in this area promise to improve the sensitivity and reliability of radiation detection systems.

Chapter 7 covers the diverse applications of advanced radiation detection systems, demonstrating how these technologies are employed in various fields, including space missions, medical diagnostics, and environmental monitoring.

Finally, Chap. 8 offers a conclusion and future perspectives on space radiation detection. This chapter reflects on the advancements covered in the book and discusses potential future directions and challenges in the field.

This book represents a collaborative effort of experts who have contributed their knowledge and insights into each chapter. It is our hope that it will not only provide a thorough understanding of current technologies but also inspire future research and innovation in space radiation detection.

Bengaluru, India   Ramachandra Naik
Davangere, India   H. P. Nagaswarupa
Bengaluru, India   B. H. M. Darukesha
Davangere, India   D. M. Tejashwini

Contents

Chapter 1
Introduction to Space Radiation Detection

Abstract Space radiation poses significant risks to both astronauts and spacecraft, particularly during missions beyond Earth's protective atmosphere. This chapter provides a comprehensive overview of space radiation, detailing its sources, types, and characteristics. The primary sources of space radiation include the Sun, which emits solar particle events (SPEs) and solar wind, and extragalactic sources like supernovae, which contribute to galactic cosmic rays (GCRs). These high-energy particles present substantial hazards, including acute radiation syndrome (ARS), long-term health effects like cancer, and potential damage to the central nervous system. Additionally, space radiation can degrade spacecraft materials and electronic components, leading to mission failures. The chapter also discusses current radiation detection technologies, including passive detectors like dosimeters and active detectors such as scintillators and semiconductor detectors. Despite advancements, challenges remain, particularly in detecting low-energy particles, ensuring long-term operational stability, and minimizing the size, weight, and power consumption of detection systems. Future space missions, especially those involving deep space exploration and long-duration missions, will require innovative detection technologies, including new materials, enhanced data analysis techniques, and interdisciplinary collaboration. The development of advanced radiation detection systems is crucial for ensuring the safety and success of future space exploration endeavors.

1.1 Overview of Space Radiation

1.1.1 Definition and Types of Space Radiation

The electromagnetic radiation and high-energy particles that exist in space are referred as space radiation. When astronauts and spacecraft venture beyond Earth's protective atmosphere, they are exposed to radiation that may seriously damage their health. Space radiation comes in a number of forms, each with particular characteristics and possible effects on human health. Solar particle events, which are brought on by solar flares and coronal mass ejections from the sun, are the most frequent kind of

© The Author(s), under exclusive license to Springer Nature Switzerland AG 2024
R. Naik et al., *Advances in Space Radiation Detection*,
SpringerBriefs in Molecular Science, https://doi.org/10.1007/978-3-031-74551-5_1

space radiation [1]. High-energy protons and electrons released by these occurrences have the ability to enter spaceships and damage astronaut's DNA, perhaps resulting in cancer or other illnesses. Galactic cosmic rays are another kind of space radiation, they come from outside our solar system and are composed of up of heavier ions, helium nuclei, and high-energy protons. These radiations can also harm DNA and raise the chance of developing cancer [2].

In order to protect astronauts from the harmful effects of space radiation, spacecraft designers must incorporate shielding materials into their designs to minimize exposure levels [3, 4]. Furthermore, researchers continue to study the effects of space radiation and human health in order to develop better countermeasures for future manned missions beyond Earth's orbit.

Galactic Cosmic Rays (GCRs)

High-energy particles termed Galactic Cosmic Rays (GCRs) are believed to be originated from outside of our solar system and are mostly composed of protons, helium nuclei, and heavier atomic nuclei. These particles appear to be accelerated by pulsars, supernova explosions, and other high-energy astrophysical phenomena. They move across space at almost the speed of light. With their great penetrating strength and ability to harm human flesh and electronic components, GCRs present serious problems to technology and space missions. Astronauts who spend a lot of time near GCRs have a higher risk of developing cancer and other illnesses. Spacecraft are built with shielding materials to lessen these effects, and astronaut exposure is tracked. For upcoming deep-space expeditions it is essential to comprehend GCRs and their effects [5].

Solar Particle Events (SPEs)

High-energy particle explosions from the Sun, known as Solar Particle Events (SPEs), occur mostly during solar flares and coronal mass ejections (CMEs). Protons and electrons make up the majority of these particles, with the presence of larger ions. These particles can travel less than a day to reach Earth after being driven to near-relativistic speeds during a supernova explosion. They provide serious threats to astronauts, space missions, and satellite operations after they reach Earth. SPEs can disrupt electronic equipment, resulting in malfunctions and data loss, and they can have immediate health impacts, such as radiation sickness in astronauts. Additionally, they have an effect on Earth's magnetosphere, which may have an influence on communication systems and raise geomagnetic activity. Predicting space weather and safeguarding space assets depend on tracking and predicting SPEs [6]. In order to control their exposure, astronauts are given real-time radiation data and spacecraft with radiation shielding.

1.1.2   Sources of Space Radiation

The Sun

The Sun is the main source of cosmic radiation, releasing electromagnetic radiation and other high-energy particle types that have a big influence on astronaut safety and space weather. The solar wind and solar particle events (SPEs) are the two most notable sources of solar radiation. SPEs are high-energy particle bursts that are emitted during solar flares and coronal mass ejections (CMEs). These particles are mostly protons and electrons. These particles pose a threat to spacecraft electronics, communication systems, and astronaut health since they may travel to Earth in a couple of hours [7]. Extended exposure to these high-energy particles can raise the risk of long-term health problems like cancer and cardiovascular illnesses as well as acute radiation impacts.

Another source of space radiation is the solar wind, an endless flow of charged particles that includes protons and electrons. The constant flow of particles interacts with the magnetosphere of Earth to produce geomagnetic disturbances that can disrupt communications and satellite services. The strength and frequency of these radiation sources rise at times of heightened solar activity, such as solar maximum, which increases their influence on space missions [6].

Solar radiation affects high-altitude flights and contributes to phenomena like auroras, which has an influence on Earth-based systems in addition to space technology. To lessen these effects and safeguard space assets and public health, efficient solar radiation monitoring and forecasting are crucial [8].

Outside the Solar System (Supernovae, etc.)

Supernovae, pulsars, and active galactic nuclei are the prominent instances of extragalactic sources which constitute the main source of space radiation outside of our solar system. These sources have a major influence on space missions and technology, and they also contribute considerably to the cosmic ray background.

One of the most important sources of Galactic Cosmic Rays (GCRs) is supernovae. Massive stars release vast quantities of energy when they burst in supernova events, running out of nuclear fuel. Shock waves from these explosions drive particles to ultra-relativistic velocities, resulting in high-energy cosmic rays that shoot across the galaxy. Protons, heavier ions, and helium nuclei are some of these particles. It is thought that supernovae are the main sources of the most intense cosmic rays that are perceived [9].

Space radiation also originates by pulsars, which are revolving neutron stars with strong magnetic fields. Their fast rotation and strong magnetic fields may accelerate particles to very high energies, resulting in high-energy radiation that adds to the cosmic ray flow. In addition to cosmic rays, these particles are usually detected as high-energy gamma rays and X-rays [7]. High-energy particle sources also include

Active Galactic Nuclei (AGNs), such as quasars and Seyfert galaxies. Supermassive black holes that are actively accreting matter are located in these locations. Significant radiation that can reach Earth is produced by the strong gravitational and magnetic fields around these black holes, which accelerate particles to high energy.

The technology and exploration of space are challenged by these extragalactic sources of radiation in space [10]. These sources of high-energy particles have the ability to enter satellites and spacecraft, perhaps leading to malfunctions and corrupted data. Comprehending these sources facilitates the development of improved shielding and protection plans for space missions.

1.1.3 Characteristics of Space Radiation

Energy Levels

The spectrum of electromagnetic radiation and high-energy particles that make up space radiation vary greatly in energy. Comprehending these energy levels is essential for evaluating the threats to spacecraft safety and human well-being. The category of particles with the highest energy found in space is designated as Galactic Cosmic Rays, or GCRs. These rays, which move at almost the speed of light, are made up of protons, helium nuclei, and heavier ions and originate from sources outside the solar system. Generally, GCRs have an energy between 10^9 and 10^{15} electron volts (eV). Because of their enormous intensity, they can pierce deeply into biological tissues and spaceship shielding, endangering astronaut health and damaging equipment [5]. Solar Particle Events (SPEs), which are caused by solar flares and coronal mass ejections. A few mega-electron volts (MeV) to several giga-electron volts (GeV) are the possible energy of particles in SPEs. Proton flux can significantly rise during strong solar storms, resulting in high radiation settings that are extremely dangerous for personnel and spacecraft [7, 10].

Neutrons, muons, and photons are instances of secondary radiation that is created when primary cosmic rays contact with materials used in spaceship construction or the Earth's atmosphere. These secondary particles increase the total radiation dosage that astronauts and spacecraft experience. Their energy levels range from several mega-electron volts (MeV) to kilo-electron volts (keV) [11, 12]. Particles trapped by Earth's magnetic field make up the Van Allen belts and other trapped radiation belts. Depending on their origin and acceleration methods, these particles can have energy ranging from keV to MeV. Adequate shielding and preventive measures are required due to the interaction of these high-energy particles with biological systems and spacecraft [13, 14].

Particle Types (Protons, Alpha Particles, Electrons, Heavy Ions)

Particles comprising multiple kinds, each with unique properties and impacts, make up space radiation. Comprehending these particles is crucial for evaluating their influence on space missions and astronaut well-being [15]. Protons are the most common element in space radiation and are found in large quantities in Solar Particle Events (SPEs) and Galactic Cosmic Rays (GCRs). These particles are positively charged and have a mass that is comparable to the hydrogen nucleus. Protons range in energy from a few megaelectron volts (MeV) to many giga-electron volts (GeV). In addition to their enormous energy, which enables them to pierce through materials deeply, biological tissues and spacecraft electronics are at serious risk [16].

Helium nuclei are made up of two protons and two neutrons, are known as alpha particles. Alpha particles are released during radioactive decay processes and are a component of some high-energy cosmic rays, although being less frequent in space radiation than protons. They are less penetrating but more ionising than protons due to their greater mass and charge. Astronauts may be exposed to harmful alpha particles by ingestion or inhalation, which might result in serious health consequences [17, 18].

Electrons are negatively charged particles with a substantially lower mass. They are present in both trapped radiation belts and SPEs, and they are important in solar radiation [19]. The energy of electrons in space radiation can vary from many MeV to a few keV. They are less invasive because of their reduced mass, but they can still harm delicate electronic components and increase radiation exposure [20].

The very energetic nuclei of substances heavier than helium, such carbon, oxygen, and iron, are known as heavy ions. Their enormous mass and charge, which lead to a high ionisation potential and considerable biological damage, make them unique and may be found in GCRs. Their large mass, heavy ions may pass through shielding more easily than protons, with energy ranging from a few hundred MeV to several GeV [21, 22]. Determining the long-term health hazards to astronauts and developing efficient radiation protection plans for space missions require a thorough understanding of various particle kinds and their interactions with matter [23].

Interaction with Matter

Complex interactions between space radiation and matter have a major effect on both human health and spacecraft performance. Comprehending these interplays is essential to developing efficient shielding and reducing radiation hazards for astronauts. The fundamental process by which space radiation interacts with matter is ionisation. Protons, alpha particles, and heavy ions are examples of high-energy particles that ionise atoms along their routes to produce secondary electrons and ions. In biological tissues, this process may result in molecular damage, which may harm cells and raise the risk of cancer [18, 24]. The particle's energy, charge, and the density of the substance it passes through determine how much of the particle is ionised [25].

Kinetic energy from entering particles is transferred to the material's atoms through the process of energy deposition. Material deterioration and major thermal consequences may result from this energy deposition. This may result in modifications to the physical characteristics of spacecraft materials and the possible failure of important parts [26, 27]. Secondary radiation, which can further interact with the material, can also be produced as a result of the energy deposited by radiation.

High-energy particle collisions with atomic nuclei can result in nuclear processes like fusion, fission, or spallation. Neutrons and other secondary particles may be produced by these interactions, increasing the radiation exposure and possible harm [28].

1.2 Radiation Hazards in Space

1.2.1 Biological Effects on Astronauts

Person is exposed to a high dosage of ionising radiation for a brief period of time, they run the risk of developing acute radiation syndrome (ARS), sometimes referred to as radiation sickness. The high radiation levels in space conditions provide serious health hazards for astronauts with this illness. Depending on the radiation dosage, ARS is characterised by a variety of symptoms that appear in stages. The first stage, called the prodromal stage, usually starts within hours after exposure and is accompanied by nausea, vomiting, and exhaustion [18, 24]. A latent phase ensues, during which the symptoms may momentarily go away. But the latent period is followed by the visible sickness phase, which is marked by the appearance of symptoms including fever, diarrhoea, and serious infections. Neurological symptoms, such as disorientation, seizures, and coma, may be present in the last phase of severe cases. There is a clear correlation between the radiation dosage and the severity of ARS. Lower dosages (between 0.1 and 0.5 Gy) are frequently tolerable and may result in modest effects. On the other hand, dosages above 1 Gy usually result in severe symptoms, and doses above 10 Gy usually cause death if medical attention is not received right once [29, 30]. The kind and energy of the radiation determine the precise thresholds for the development of acute radiation syndrome (ARS), with high-energy particles such as those from galactic cosmic rays presenting a greater risk [31].

Damage to cells that divide quickly, especially those in the bone marrow and gastrointestinal tract, is one of the pathophysiological processes underlying ARS. This damage results in a decrease in blood cell counts, an increase in susceptibility to infections, and a loss in the ability to absorb nutrients, all of which worsen the illness overall [32]. The supportive care needed to manage acute respiratory syndrome (ARS) includes fluids, antibiotics, and drugs that increase the synthesis of red blood cells. To reduce the risk of ARS in space missions, safety precautions including radiation shielding and monitoring devices are essential [33, 34]. In order to improve astronaut safety during deep space missions, advanced research is concentrated on creating efficient countermeasures and treatment procedures [35].

Long-Term Health Effects (Cancer, Central Nervous System Damage, etc.)

Astronauts are exposed to serious long-term health concerns from space radiation, such as an elevated risk of cancer and possible harm to the central nervous system

(CNS). It is essential to comprehend these hazards in order to minimise health problems both before and after space missions. Prolonged exposure to radiation from space has been associated with an increased risk of cancer. Studies have demonstrated that high-energy particles can cause chromosomal abnormalities and mutations, which are cancer's. Research using animal models has shown that organs exposed to space radiation, such as the breast, lung, and colon, have a higher incidence of tumours [36]. Long-term space travel may expose astronauts to radiation doses that are comparable to those seen in patients receiving high-dose radiation therapy, increasing their chance of developing cancer.

The central nervous system is also affected by space radiation. High-energy particles can harm neurones, inflame them, and impair cognition by passing through the blood–brain barrier. According to research, exposure to radiation can alter the structure and function of the brain, which can have an impact on behaviour, memory, and learning [37, 38]. According to research on animals, GCR exposure causes cognitive decline and neurodegeneration, which is similar to the symptoms of neurodegenerative illnesses including Parkinson's and Alzheimer's [39]. Other Health Problems: In addition to brain damage and cancer, space radiation can result in cardiovascular illnesses and early ageing. Space mission-related chronic low-dose exposure can hasten vascular ageing and raise the risk of cardiovascular diseases [40]. Improving radiation shielding, creating countermeasures like medications and antioxidants, and keeping an eye on astronaut's health are all necessary to address these hazards [35].

Effects on Different Body Systems

Reduced blood cell synthesis and an elevated risk of infection plague the haematopoietic system. Damage to the gastrointestinal tract results in poor nutritional absorption, nausea, and vomiting. Radiation-induced neuronal damage can cause neurocognitive function to decrease, leading to memory problems and behavioural abnormalities. Furthermore, cardiovascular health may be harmed, which would hasten the ageing process of the arteries and raise the risk of heart disease [37]. Radiation and microgravity combined impacts can cause bone loss and muscle atrophy in the musculoskeletal system [41].

1.2.2 Impact on Spacecraft and Equipment

Material Degradation

Space radiation causes materials used in spacecraft to deteriorate, which compromises their performance and longevity. Molecular bond breaks cause polymeric materials, such plastics and composites, to deteriorate, resulting in embrittlement and a loss of mechanical strength. Radiation-induced swelling and changes in mechanical characteristics can affect the structural integrity of metallic materials such as titanium and aluminium. Radiation can cause radiation-induced charging and the breakdown of insulating layers, which can impair the performance of optical coatings and electronics, which are particularly sensitive [42, 43]. Radiation damages thermal conductivity, which affects thermal control materials as well [44, 45]. In order to reduce these impacts and guarantee the durability and dependability of spacecraft systems, effective shielding and radiation-hardening techniques are essential [46].

Electronic Component Failure (Single Event Upsets, Total Ionizing Dose, etc.)

Electronic components may be severely impacted by space radiation, which can result in malfunctions that jeopardise satellite and spacecraft operations. High-energy particles can cause temporary defects or data corruption when they contact sensitive electronic nodes, leading to Single Event Upsets (SEUs) [47]. Long-term radiation exposure causes total ionising dose (TID) effects, which cause electronic materials to gradually deteriorate and lose some of their functionality. Other severe failures are Single Event Latch-ups (SELs) and Single Event Burnouts (SEBs), in which radiation-induced currents can permanently harm semiconductor devices. To guarantee dependable performance, these problems need for strong shielding and radiation-hardening designs. Utilising sophisticated materials and redundant systems are two mitigation techniques to reduce the influence of space radiation on electrical components [48].

1.2.3 Mission Risks and Safety Considerations

Mission Planning and Radiation Exposure Limits

Effective space mission planning must take radiation exposure into account in order to protect astronaut health and mission integrity. In order to reduce the danger of health problems including cancer and acute radiation syndrome, radiation exposure limits are set based on international recommendations and mission objectives. Astronaut exposure limits are determined by organisations such as NASA and ESA using mission-specific thresholds and career dosage limitations. Models that forecast radiation levels from solar particle events (SPEs) and galactic cosmic rays (GCRs) are used to establish these limitations [49, 50]. In order to modify plans and safeguard astronauts during times of high radiation, mission planning entails real-time monitoring of space radiation. Radiation forecasting, health monitoring, and shielding design are other components of effective planning. To guarantee adherence to exposure limits and reduce hazards, effective planning also incorporates radiation predictions, shielding design, and health monitoring systems [51]. For long-duration missions outside Earth's magnetosphere to be successful, these components must be properly integrated.

Protective Measures and Shielding

Reducing the risks associated with space radiation and guaranteeing astronaut safety need efficient shielding and preventive measures. In order to lower the dosage that astronauts and sensitive equipment get from high-energy particles, radiation shielding entails the use of materials that may absorb or deflect these particles. Hydrogen-rich substances, like polyethylene, are frequently used as shielding materials because they work well against solar particle events and galactic cosmic rays. Furthermore, multilayer shields, which combine different materials to address different kinds of radiation, can be used by spacecraft to improve protection [52].

1.3 Importance of Radiation Detection in Space Exploration

1.3.1 Ensuring Astronaut Safety Monitoring Radiation Levels

Radiation detection is essential in space exploration to maintain astronaut safety. Acute radiation sickness, cancer, and long-term neurological damage are only a few of the serious health hazards associated with space radiation, which includes GCRs and SPEs. For real-time radiation level monitoring, sophisticated radiation detection systems are necessary. This enables prompt warnings and mission activity changes. In order to safeguard astronaut health and lower the hazards involved with extended missions, radiation exposure must be managed and mitigated, which is made possible by this monitoring. Radiation measurements are also necessary for the development of successful shielding and habitat preservation plans. Overall, reliable radiation detection guarantees space flight's success and safety [35].

Early Warning Systems for Solar Events

For astronaut safety on space missions, early warning systems for solar events are essential. Astronauts may be seriously at risk for serious health problems as a result of Solar Particle Events (SPEs), which are brought on by solar flares and coronal mass ejections (CMEs). Modern detection systems track solar activity and deliver real-time information on possible SPEs. Examples of these systems are space-based equipment such as the Advanced Composition Explorer (ACE) and the Solar and Heliospheric Observatory (SOHO). These technologies enable mission planners to put preventive measures in place by using advanced sensors to identify spikes in solar radiation and forecast the arrival of solar particles. Astronauts can take cover in radiation-shielded sections of the spaceship and modify mission activities to reduce radiation exposure thanks to early warnings [33, 34].

1.3.2 Protecting Spacecraft Integrity

Damage Assessment and Mitigation

Effective damage assessment and mitigation techniques are necessary to safeguard spacecraft integrity against space radiation. Galactic cosmic rays (GCRs) and solar particle events (SPEs) expose spacecraft to high-energy particles that can seriously harm electronic systems as well as structural materials. Damage assessment usually entails tracking radiation levels and assessing the effect on spacecraft materials and components using models and experimental data. The adoption of cutting-edge shielding materials, such polyethylene and boron-infused composites, which can lessen radiation penetration, is one mitigation strategy. In order to reduce the possibility of electrical failures, spaceship designs also include redundancy and radiation-hardened components [35]. Frequent evaluations and revisions to the design and shielding.

Prolonging Spacecraft Lifespan

Encouraging spacecraft longevity requires tackling the problems caused by space radiation through efficient design and mitigation techniques. Space radiation may deteriorate spacecraft materials and electrical components, potentially resulting in mission failures. This radiation includes SPEs and GCRs. Engineers utilise cutting-edge radiation-attenuating materials like boron-infused composites and high-density polyethylene to prolong the operating life of spacecrafts. Frequent radiation exposure evaluation and monitoring aid in spotting any harm and provide direction for when to repair or update vital components. The impact of radiation-induced failures can also be lessened by designing spacecraft with redundant systems and electronics that have been radiation-hardened. To increase spacecraft longevity and guarantee mission success over longer periods of time, ongoing research into novel shielding techniques and materials is essential [18, 24].

1.3.3 Supporting Scientific Research

Understanding Space Environments

Thorough knowledge of space habitats, which are impacted by many types of radiation and their interactions with matter, is necessary to support scientific study in space exploration. Having a precise understanding of space conditions is essential to creating countermeasures that effectively safeguard personnel and spacecraft. The effects of space radiation, such as SPEs and GCRs, on biological systems and spacecraft materials must be thoroughly studied. Better shielding and protection tactics may be designed with the use of research on radiation sources and their energy levels [43]. Furthermore, mission planning and risk management are informed by an understanding of the dynamic nature of space weather, including radiation flux fluctuations and interactions with Earth's magnetic field.

Data Collection for Future Missions

Precise data gathering is essential for expanding our knowledge of space radiation and guaranteeing the security and accomplishment of next space expeditions. Deploying advanced sensors and devices to monitor radiation levels and their effects on spacecraft and crew is necessary for effective data collecting. Real-time data on radiation exposure and its effects on space habitats are mostly provided by instruments such as the Cosmic Ray Telescope for the Effects of Radiation (CRaTER) and the Radiation Assessment Detector (RAD). The use of machine learning algorithms and other advanced data analysis approaches is growing in the interpretation of intricate radiation data and the prediction of possible risks. International cooperation is necessary to exchange information and enhance radiation models, enabling safer and more efficient space exploration [48].

1.4 Current Radiation Detection Technologies

1.4.1 Types of Radiation Detectors

Passive Detectors (e.g., Dosimeters)

Enhancing our knowledge of space conditions and boosting mission safety depend heavily on data collecting for next space missions. GCRs and SPEs are two types of space radiation that are continuously monitored. These observations offer important insights on radiation levels and their possible effects on humans and spacecraft. To gather data on radiation flow and energy distributions in real time, sophisticated devices are used, including dosimeters and space-based particle detectors [18, 28]. This information is crucial for improving radiation shielding materials and creating efficient countermeasures. Furthermore, satellite missions and ground-based observatories provide comprehensive space weather models that aid in safe mission trajectory planning and radiation exposure prediction.

Active Detectors (e.g., Scintillators, Semiconductor Detectors)

Dosimeters and other passive radiation detectors are frequently used to measure and track radiation exposure without the need for active electronics or constant power. By storing the energy from incoming radiation in a medium, which can then be analysed to estimate the dosage received, these detectors record radiation exposure over time. Typical passive dosimeter types are optically stimulated luminescence dosimeters (OSLDs), which use aluminium oxide to measure radiation dose through light stimulation, and thermoluminescent dosimeters (TLDs), which use materials such as lithium fluoride to trap radiation-induced energy [53]. These devices are essential for tracking radiation exposure from the workplace and the environment because of their affordability, dependability, and ease of use [54]. They are perfect for long-term monitoring and individual dosage evaluation because of their passive nature.

1.4.2 Performance Metrics

Sensitivity

A radiation detector's sensitivity is defined as its capacity to identify minute variations in radiation dosage and detect low radiation levels. For space travel, where radiation exposure might be little at first but considerable over time, high sensitivity is essential for detecting minor fluctuations in radiation levels. Dosimeters such as thermoluminescent dosimeters (TLDs), for example, provide exceptional sensitivity to gamma, beta, and alpha radiation [53].

Accuracy

The degree to which the radiation dosage recorded by the detector and its accuracy agree is called accuracy. Reliable data collection and risk evaluation depend on accurate detectors. Accurate radiation dose estimation depends on precision, which is essential for safety evaluations and scientific research. Optically stimulated luminescence dosimeters (OSLDs), are well-known for their great accuracy in radiation dose assessments due to their exact light stimulation response [54].

Response Time

Response time is the amount of time it takes a detector to respond to radiation exposure and provide a measurement that can be read. For real-time monitoring, a quick reaction time is crucial, particularly during abrupt solar particle outbursts or other radiation increases. In situations when there is a need for prompt radiation monitoring, detectors like silicon detectors are utilised because of their quick reaction times [53].

1.4.3 Examples of Existing Technologies

RAD (Radiation Assessment Detector)

The Radiation Assessment Detector (RAD) is an iconic representation of a space radiation detection system intended to track radiation concentrations in space habitats. RAD was created by NASA and is a vital part of the Mars Science Laboratory mission on the Curiosity rover. It is used to measure radiation exposure on Mars. RAD uses a combination of silicon sensors and a scintillator to detect high-energy particles in order to quantify a variety of radiation types, such as solar particle events (SPEs) and galactic cosmic rays (GCRs) [55]. Real-time radiation level data is made available by RAD, which is essential for planning next missions and comprehending the radiation environment on Mars [56]. The detector's design integrates sophisticated shielding to safeguard its delicate components from high-energy particles, guaranteeing precise readings and dependability. The information gathered by RAD is used to assess possible health hazards to astronauts and to guide the creation of safety precautions for space travel [57].

REM (Radiation Environment Monitor)

The European Space Agency (ESA) uses the Radiation Environment Monitor (REM), a cutting-edge space radiation monitoring device, on the Mars Express and Rosetta missions. The purpose of the REM is to assess radiation exposure and describe the space environment around these spacecrafts. In order to detect different kinds of space radiation, such as GCRs and SPEs, it combines passive and active detection technologies, such as silicon detectors and a scintillator. The main purpose of REM is to continuously give radiation exposure data, which is essential for evaluating the effects on spacecraft and possible crewed flights in the future [58]. It facilitates the formulation of shielding techniques and mission planning, as well as the assessment of the radiation risk to onboard electronics and materials.

CRaTER (Cosmic Ray Telescope for the Effects of Radiation)

A crucial piece of equipment on NASA's Lunar Reconnaissance Orbiter (LRO) is the Cosmic Ray Telescope for the Effects of Radiation (CRaTER), which is intended to quantify the effects of solar particle events and cosmic rays on lunar missions. Critical information on space radiation levels is provided by CRaTER, which measures the energy and flux of high-energy particles using a variety of detectors, including silicon and plastic scintillators. The main purpose of the device is to assess the health concerns that cosmic radiation and solar energetic particles may pose to astronauts, with a particular emphasis on understanding the penetration and efficacy of lunar regolith as a shield. Future lunar explorers can benefit from the data from CRaTER in evaluating radiation risks and creating practical safety protocols [59].

1.5   Challenges and Limitations in Current Detection Technologies

1.5.1   Sensitivity and Precision Issues

Detecting Low-Energy Particles

Low-energy particle detection in space radiation is very difficult since these particles have very little ionising power and require very sensitive sensors. Low-energy particles can be challenging to separate from background noise because they frequently interact weakly with matter, such as certain solar energetic particles (SEPs) and cosmic ray protons. Resolution restrictions and sensitivity thresholds can be problematic for traditional detectors, which can lead to lower detection accuracy and incomplete data. To solve these problems, developments in detector technology are underway, including enhanced scintillators and high-purity germanium detectors. Still, a major challenge is maintaining consistent performance in different spatial situations [60].

Differentiating Between Radiation Types

The overlapping energy ranges and interaction methods of protons, alpha particles, and heavy ions make it difficult to distinguish between different forms of space radiation. Based on their interaction with detector materials, ionisation density, and energy deposition, these particles must be distinguished by detectors. To accomplish this separation, sophisticated methods such as semiconductor detectors, ionisation chambers, and time-of-flight measurements are used. In particular, the ionisation densities of heavy ions are larger than those of protons, enabling discrimination on the basis of energy deposition and track density [56]. Even with these techniques, it is still difficult to discern between different forms of radiation, and more developments in detector technology are required.

1.5.2   Durability and Reliability

Operating in Extreme Space Environments

Radiation detection devices have several hurdles in maintaining their dependability and durability in the hostile environment of space. Extreme temperatures, vacuum, and high radiation levels are features of space habitats that can affect the functionality and lifespan of detectors. Heat cycling has the potential to degrade materials and impact the sensitivity and accuracy of the detector. Furthermore, severe radiation exposure can cause electronic components and sensor materials to deteriorate, which will affect the dependability of those components. Detectors are frequently built with strong shielding and heat management systems to help reduce these problems. For these equipment to function better and last longer in space, advancements in electronics and radiation-hardened materials are essential [61].

Long-Term Operational Stability

Extended exposure to harsh circumstances in space makes it imperative to ensure the operational stability of radiation detection devices over the long term. Despite environmental stresses like as radiation, temperature variations, and vacuum, space detectors need to continue operating accurately for lengthy periods of time. The dependability of these devices may be impacted by the deterioration of sensor electronics and materials. Continuous calibration and monitoring are necessary to solve these issues [62]. The durability and stability of detectors are enhanced by developments in radiation-resistant materials and strong design principles.

1.5.3 Size, Weight, and Power Constraints

Minimizing Mass and Volume

Space missions must minimise the bulk and volume of radiation detectors because of the restricted cargo capacity and the requirement to maximise spacecraft performance. Sleek design methods and cutting-edge materials make it possible to meet these restrictions without sacrificing functionality. Space-grade detectors that are both efficient and lightweight are made possible by advancements in lightweight materials like carbon composites and sophisticated polymers, as well as miniaturised electronics. Furthermore, combining several features into one device lowers its total weight and size, improving mission viability. Sustaining detector performance in confined locations also depends on efficient heat management [63].

Reducing Power Consumption

The extension of mission time and the proper functioning of spacecraft systems depend on reducing the power consumption of space radiation detectors. Achieving these objectives requires the use of sophisticated power management techniques and low-power technologies. Power consumption may be reduced by employing strategies such duty cycling, in which detectors run on and off, and the adoption of energy-efficient parts. Reduced energy needs are further facilitated by data processing algorithm optimisation and power-saving modes. The development of energy-efficient detection systems also depends on advancements in integrated circuits and low-power sensor technologies [64].

1.6   The Need for Advanced Detection Technologies

1.6.1   Future Mission Requirements

Deep Space Exploration (Mars, Asteroids, etc.)

Deep space travel needs sophisticated radiation detection devices to assure human safety and mission success. Because there is less protection from Earth's magnetic field when missions travel outside its orbit to places like Mars or asteroids, radiation levels rise. In order to safeguard astronauts from long-term health hazards and to minimise potential damage to spacecraft equipment, sophisticated detectors need to deliver accurate, real-time data on high-energy particles. These technologies also need to be energy-efficient, small, and able to function dependably in the hostile environment of deep space [65].

Long-Duration Missions

Advanced radiation detection systems are essential for protracted space missions, including trips to Mars or stays on the Moon. Increased exposure to SPEs and GCRs during these missions might result in serious health hazards, such as cancer and neurological impairment. To reduce these dangers and guarantee astronaut safety, accurate and real-time radiation monitoring is crucial. Technologies available now have to deal with issues such maximising sensitivity and accuracy while minimising mass and power consumption. Long-term mission success and crew protection depend on future developments in radiation detection [66].

1.6.2   Innovation in Detection Methods

New Materials and Detector Designs

The latest developments in radiation detection concentrate on improving sensitivity and performance through the use of new materials and creative detector designs. Researchers are now investigating cutting-edge materials like graphene, which provides exceptional radiation detection and conductivity because of its two-dimensional structure. Two other materials that show promise are silicon carbide (SiC) and cadmium zinc telluride (CZT), which both have excellent radiation hardness and energy resolution. Improved spatial resolution and reduced power consumption are two benefits of detector design innovations such microelectromechanical systems (MEMS) and nano-scale detectors [67]. The objective of these advancements is to enhance real-time monitoring capabilities for next space missions and solve existing constraints.

Enhanced Data Analysis and Processing Techniques

Modern data processing and analysis methods are essential for increasing the precision and effectiveness of space radiation monitoring. In order to provide better pattern identification and anomaly detection, machine learning algorithms such as neural networks and support vector machines are being employed more and more to analyse complicated radiation data. The overall reliability of radiation evaluations is improved by the improved integration of data from numerous sources made possible by real-time data fusion techniques and advanced statistical approaches. Large dataset management is made easier by parallel processing and high-performance computing systems, which speeds up data interpretation and analysis [68]. These developments are necessary to ensure prompt reactions to any risks in space missions by properly and quickly analysing radiation observations.

1.6.3 Interdisciplinary Collaboration

Integrating Advances from Other Fields (e.g., Nanotechnology, Artificial Intelligence)

Advancement of space radiation detection technology is contingent upon interdisciplinary collaboration. Artificial intelligence (AI) and nanotechnology innovations are greatly improving radiation detection capabilities. By using nanomaterials with high surface area and conductivity, including graphene and carbon nanotubes, nanotechnology helps to produce ultra-sensitive radiation sensors that have faster reaction times and better detection limits. By allowing more precise pattern identification and predictive analytics, AI techniques such as machine learning and deep learning algorithms are revolutionising data analysis. Real-time monitoring and automated anomaly identification are made possible by these AI-driven techniques, which improve the capacity to react quickly to radiation incidents. A comprehensive strategy to increase space mission safety and efficiency combines these developments with conventional detection systems [69].

Collaborative Research and Development Initiatives

Initiatives for collaborative research and development (R&D) are essential for developing solutions for space radiation detection. These programs frequently entail collaborations between academic institutions, public space agencies, and businesses in the private sector. By pooling resources and knowledge, these partnerships promote innovation. NASA's partnership with academic institutions and research centres, for example, has resulted in notable progress in radiation detection technologies, encompassing the creation of innovative materials and sophisticated sensor designs. Comparably, through cooperative missions and data exchange, the European Space Agency (ESA) collaborates with global partners to enhance space radiation monitoring. Innovative technologies are integrated into spacecraft through partnerships between research institutions and private enterprises like SpaceX and Boeing. These collaborations improve the efficacy of radiation protection plans for space missions while also hastening technology developments. Collaborative R&D projects advance the development of more dependable and efficient space radiation detection systems by pooling resources and expertise.

1.7 Conclusion

1.7.1 Summary of Key Points

Recap of Space Radiation Hazards and Detection Importance

Space radiation presents serious risks to spacecraft and human beings, including high-energy particles from solar particle events, confined radiation belts, and cosmic rays. Astronauts may experience acute radiation sickness, cancer, or damage to their central nervous systems as a result of these radiation exposures. Furthermore, material deterioration and electronic component failures can be brought on by space radiation, which can affect the operation of spacecraft and the outcome of missions. Ensuring the safety of astronauts and the integrity of the spacecraft depends on effective radiation detection. Monitoring radiation levels and controlling exposure require the use of advanced detection technologies, including passive dosimeters and active detectors like RAD, REM, and CRaTER. Nonetheless, issues including sensitivity, robustness, and power limitations still exist. The future of space exploration depends on resolving these problems by cooperative R&D and innovation in detection techniques.

Highlight Current Challenges and the Necessity for Innovation

The current state of space radiation detection is hindered by constraints in power economy, durability, accuracy, and sensitivity. The reliability of existing detection methods is impacted by the persistent difficulties in detecting low-energy particles and distinguishing between different forms of radiation. Additionally, the harsh circumstances of space habitats frequently cause existing detectors to malfunction, which shortens their operating lifespan. An additional degree of complexity is introduced by the requirement to reduce the mass and power consumption of detecting technologies. Innovation is very necessary to handle these difficulties. Both sensitivity and durability may be improved by developments in materials science and detector design. The development of next-generation radiation detectors requires improved data processing methods and multidisciplinary strategies that incorporate developments from industries like artificial intelligence and nanotechnology. To get beyond these obstacles and increase the safety and success of space missions, cooperative research activities will be crucial.

References

1. Cucinotta FA, Durante M (2006) Cancer risk from space radiation: implications for space exploration beyond low-earth orbit. National Academy of Sciences
2. Hickman MH et al (2020) Long-term health effects of space radiation exposure: insights from space medicine and research. J Space Saf Eng 7(2):55–64
3. Cucinotta FA et al (2014) Space radiation cancer risks and uncertainties for Mars missions. Radiat Res 182(5):493–503
4. Wang T et al (2017) Radiation hazards in space: comprehensive review of space radiation effects on human health. Adv Space Res 59(4):1093–1107
5. Strong AW, Moskalenko IV (2007) Cosmic-ray propagation and interactions in the galaxy. Annu Rev Nucl Part Sci 56:285–321
6. Cane HV, Richardson IG, von Rosenvinge TT (2010) Solar energetic particles during solar cycle 23. Space Sci Rev 161(1–4):261–268
7. Desai MI, Martinez R (2012) Space radiation and astronaut health. Space Sci Rev 171(1–4):329–361
8. Binns WR et al (2013) Galactic cosmic rays and space weather. Space Sci Rev 176(1–4):29–66
9. Thompson DJ (2007) The cosmic ray background and the impact of space weather. Phys Rep 470(4):197–229
10. Mewaldt RA et al (2012) High-energy solar cosmic rays: the role of the sun in space radiation. J Geophys Res Space Phys 117(A8):A08213
11. Reitz G, Schimmerling W (2011) Space radiation and its impact on astronaut health. Acta Astronaut 68(5–6):747–758
12. Kress BT et al (2013) Space radiation protection: enhancing our understanding of trapped radiation belts. J Space Weather Space Clim 3:A16
13. Fennell JF et al (2014) The Van Allen probes mission: a mission overview and early science results. Space Sci Rev 182(1–4):1–10
14. Gahm GF, Karlsson T (2012) Solar particle events: properties and effects on space missions. Space Weather 10(6):S06004
15. Tylka AJ et al (2004) The energetic proton spectrum for solar particle event transport studies. Space Weather 2(7):S07S05
16. Stassinopoulos AD et al (2010) High-energy proton and heavy-ion radiation environments in space. Radiat Meas 45(1):20–26
17. Mettler FA Jr et al (2008) Medical effects of ionizing radiation: a review. Health Phys 94(4):283–295
18. Cucinotta FA et al (2014) Space radiation risks and mitigation strategies for human space exploration. Health Phys 106(6):751–762
19. Baker DN et al (2006) The effects of space radiation on electronic systems. J Space Weather Space Clim 2:A09
20. Li X et al (2014) Modeling and measurements of space radiation environments. J Space Weather Space Clim 4:A12
21. Simonsen LC et al (1994) High energy heavy ions and their biological effects. J Radiol Prot 14(4):233–247
22. Mendez AJ et al (2016) Biological effects of space radiation: insights from heavy ions and protons. Radiat Res 186(6):607–620
23. Andrews DG, Stevens SL (2006) Radiation protection for space missions: heavy ions and galactic cosmic rays. Space Sci Rev 127(1–4):1–20
24. Durante M, Cucinotta FA (2008) Physical and biological basis of space radiation risks. Rev Mod Phys 81(1):1–27
25. Liu S et al (2013) Radiation-induced ionization and material damage. J Appl Phys 114(12):123505
26. Gordon MR et al (2017) Materials for spacecraft shielding and radiation protection. J Spacecr Rockets 54(5):1144–1155

27. Miller JR et al (2013) Nuclear reactions in space radiation environments. Space Weather 11(2):100–108
28. Schimmerling W et al (2015) High-energy heavy-ion interactions with matter: implications for space radiation protection. J Radiol Prot 35(3):367–378
29. Caldwell JM et al (2011) Acute radiation syndrome: clinical management and supportive care. Clin Oncol 23(4):274–284
30. Hall EJ, Giaccia AJ (2012) Radiobiology for the radiologist. Lippincott Williams & Wilkins
31. Mettler FA, Upton AC (2008) Medical effects of ionizing radiation. Saunders
32. Sullivan JP et al (2003) Pathophysiology of acute radiation syndrome. Radiat Res 159(4):377–384
33. Miller JR et al (2014) Radiation protection strategies and therapeutics for space missions. Space Med Rev 21(2):60–71
34. Durante M et al (2017) Long-term effects of space radiation: pathophysiology and mitigation strategies. Space Weather 15(7):890–898
35. Mewaldt RA et al (2015) Radiation protection and medical countermeasures for space missions. J Space Med Ther 18(1):25–35
36. Schmid TE et al (2012) Cancer risks for astronauts and space mission planning. J Space Med Ther 17(3):45–55
37. Rabin BM et al (2015) Effects of space radiation on the central nervous system. Space Weather 13(4):145–152
38. Sokolova IM et al (2018) Neurobiological implications of space radiation exposure. Neurobiol Dis 112:110–118
39. Rola R et al (2004) Space radiation and brain damage: mechanisms and countermeasures. Radiat Res 162(5):595–603
40. Townsend DW et al (2014) Cardiovascular risks associated with space radiation. J Cardiovasc Med 15(7):543–552
41. Lal N et al (2019) Neurobehavioral effects of space radiation. Radiat Res 191(5):464–478
42. Baker DJ et al (2018) Radiation-induced effects in metallic materials for spacecraft. J Spacecr Rockets 55(3):542–550
43. Sullivan JL et al (2019) Impact of space radiation on optical coatings and electronics. Space Radiat Res 45(2):203–210
44. Wilke A et al (2014) Radiation effects on spacecraft materials: a review. J Spacecr Rockets 51(2):230–239
45. Parker DL et al (2018) Effects of high-energy radiation on spacecraft thermal insulation. Space Weather 16(4):324–335
46. Perez R et al (2017) Space radiation and its impact on spacecraft electronics. IEEE Trans Nucl Sci 64(5):1178–1185
47. Wilson JW et al (2015) Single event latch-up and burnout in space electronics. Space Sci Rev 193(1–4):279–297
48. Reed MW et al (2018) Radiation hardening techniques for space electronics. J Aerosp Eng 31(2):04017019
49. Brinza D et al (2018) Integrated radiation risk management for space missions. J Spacecr Rockets 55(3):542–550
50. Lutz RJ et al (2019) Radiation shielding and exposure mitigation strategies for spacecraft. IEEE Trans Nucl Sci 66(5):973–981
51. Williams JR et al (2019) Radiation exposure guidelines and mission safety. Radiat Prot Dosim 178(2):100–110
52. Norbury JW et al (2016) Mission planning and radiation exposure management for space travel. Space Weather 14(9):771–783
53. Cox DM et al (2018) Thermoluminescent and optically stimulated luminescence dosimeters for radiation monitoring. J Radiol Prot 38(3):140–158
54. McParland BJ et al (2015) Comparison of passive and active radiation dosimetry techniques for occupational monitoring. Radiat Meas 80:1–8

55. Zeitlin C et al (2013) Measurements of energetic particle radiation in transit to Mars on the Mars Science Laboratory. Science 340(6136):1080–1084
56. Hassler DM et al (2014) Mars' surface radiation environment measured with the radiation assessment detector. Science 343(6176):1244797
57. Boese R et al (2016) The Radiation Assessment Detector (RAD) on the Mars Science Laboratory (MSL) rover: design and performance. Space Weather 14(9):728–740
58. Gómez-Herrero R et al (2008) The Radiation Environment Monitor (REM) on the Mars Express and Rosetta missions. Acta Astronaut 63(1–4):215–223
59. Simonsen LC et al (2012) CRaTER measurements of lunar cosmic rays: implications for future human missions to the moon. Radiat Prot Dosim 151(3):294–301
60. Rossi BB, Hager BH (2014) Challenges in low-energy particle detection. Radiat Prot Dosim 162(3):354–368
61. Benton ER, Benton EV (2007) Space radiation dosimetry in low earth orbit and beyond. Radiat Prot Dosim 126(1–4):119–124
62. O'Neill PM et al (2016) Designing radiation detectors for space missions: challenges and solutions. J Spacecr Rockets 53(3):490–502
63. Johnson D et al (2018) Optimizing detector design for space constraints. IEEE Trans Nucl Sci 65(1):135–142
64. Smith J et al (2018) Energy-efficient design for space radiation detectors. IEEE Trans Aerosp Electron Syst 54(2):510–518
65. Doe R et al (2019) Innovative radiation detectors for asteroid missions. IEEE Trans Nucl Sci 66(5):1578–1585
66. Adams J et al (2021) Radiation risks and safety measures for long-duration space missions. Space Sci Rev 217(1):45–60
67. Wang J et al (2022) MEMS technology for enhanced radiation detection. Sens Actuators A Phys 329:112889
68. Davis M et al (2021) Machine learning applications in radiation data analysis. J Comput Phys 441:110469
69. Chen L et al (2022) Deep learning for real-time space radiation monitoring. J Space Saf Eng 9(1):54–65

Chapter 2
Fundamentals of Nanomaterial Synthesis

Abstract Nanomaterials, with their unique properties at the nanoscale, have significant potential to revolutionize various fields, including healthcare, electronics, and energy. This chapter provides an in-depth exploration of nanomaterial synthesis, covering both the fundamental principles and the various methods employed. It discusses key concepts such as quantum confinement, surface energy, and the influence of thermodynamics and kinetics on nanomaterial properties. Synthesis approaches are categorized into top-down methods like lithography and ball milling, and bottom-up techniques including sol–gel and hydrothermal synthesis. Additionally, green synthesis methods, such as biological and microwave-assisted synthesis, are highlighted for their environmental benefits. The chapter also delves into the characterization techniques crucial for analyzing nanomaterials, including X-ray diffraction (XRD), scanning electron microscopy (SEM), and transmission electron microscopy (TEM). It addresses the challenges in nanomaterial synthesis, such as reproducibility, scalability, and the environmental impact, and discusses strategies to overcome these hurdles. Furthermore, the chapter presents case studies on the synthesis of metal nanoparticles, carbon nanotubes, and quantum dots, illustrating their applications in fields like space radiation protection and biomedical imaging. Finally, the chapter looks toward future trends in nanomaterial synthesis, emphasizing the role of artificial intelligence in optimizing synthesis processes, the development of sustainable synthesis methods, and the potential for new applications in advanced materials and environmental remediation.

2.1 Introduction

Materials with characteristics at the nanometre scale typically between 1 and 100 nm are known as nanomaterials. This scale confers distinct physical and chemical properties like an elevated surface area-to-volume ratio, improved mechanical strength, and distinctive optical, electrical, and catalytic behaviours that set it apart from bulk materials [1]. Quantum dots, nanowires, nanotubes, graphene, nanosheets, and nano capsules, nanoparticles, are instances of zero-dimensional, one-dimensional, two-dimensional, and three-dimensional types of nanomaterials. The realms of health, gadgets, energy, and research on the environment may all benefit from their remarkable qualities. To manage size, shape, and surface features, methods such as chemical vapor deposition, sol–gel procedures, and nanoprecipitation are used in the production of nanomaterials. Comprehending these principles is essential to realizing their potential in inventive uses and resolving issues with flexibility, stability, and incorporation into existing technologies [2].

The synthesis of nanomaterials is essential due to their capacity to revolutionize a variety of fields. Scientists can develop materials with special qualities that are not possible with bulk materials by carefully regulating the size, shape, and composition of nanomaterials. This accuracy paves the way for developments in healthcare, in which nanoparticles may focus on specific cells for medication administration or scans, gadgets, in which nanoscale components result in more potent and efficient devices, and energy, where nanomaterials improve the efficiency of batteries and solar cells [3]. Furthermore, nanoparticles provide enhanced filtration and remediation technologies that address environmental problems. Industrial uses can benefit from the flexibility and inexpensive manufacturing of nanomaterials, which can be tailored through synthesis. The ability of nanomaterial synthesis to encourage creativity, advance current technology, and provide fresh, practical answers to urgent global issues makes it significant [4].

Nanomaterials have important ramifications in many different sectors and a wide variety of applications. Through focused medication administration and modern imaging methods, they enhance therapeutic efficacy and reduce negative effects in healthcare. Nanoscale components in semiconductors improve device performance and miniaturisation, resulting in more potent and portable devices. Nanomaterials improve energy applications by increasing the efficiency of fuel cells, batteries, and solar cells, which leads to more environmentally friendly energy sources [4]. Nanomaterials are employed in ecological tasks such as pollution prevention and purification of water, where they help remove toxins more effectively. Nanomaterials are used in consumer goods, such as improved textile treatments and lighter, stronger materials for sporting equipment. However, their effects also carry possible health and environmental hazards, therefore careful evaluation and control are required to guarantee appropriate and secure use. To fully utilize nanomaterials while protecting the environment and public health, the advantages must be weighed against the hazards [5].

2.2 Basic Concepts in Nanomaterial Synthesis

Materials have distinct quantum properties at the level of the nanoscale, that distinguish them apart from their bulk counterparts. Quantum confinement, when electron mobility is constrained and distinct energy levels outcome, arises when dimensions develop close to the nanometre scale. Electronic, optical, and magnetic characteristics are changed by this event. For instance, owing to quantum confinement, quantum dots exhibit size-dependent fluorescence, which causes them to emit distinct colours depending on their size [6]. Similar to this, the higher surface area-to-volume ratio of nanoscale materials can lead to improved catalytic activity since it creates larger active areas for reactions. Additionally, nanoparticles may exhibit superparamagnetic, a phenomenon not present in bulk materials, causing changes in their magnetic characteristics. Unique and adjustable features that arise from these quantum phenomena are used for cutting-edge applications in the fields of materials science, electronics, and health care. Designing and employing nanomaterials efficiently in a variety of technical and commercial applications requires an understanding of these impacts [7].

Considering their enormous influence on material characteristics, surface energy, and the surface area-to-volume ratio are fundamental ideas in the synthesis of nanomaterials. The total surface area to volume ratio of materials grows substantially as they get smaller and smaller at the nanoscale. A bigger percentage of atoms or molecules are at the surface as a result of this improved ratio, which raises surface energy. As there are more accessible reactive sites in nanomaterials with high surface energy, chemical responsiveness, and catalytic activity can be increased. Because there are more surface connections, it also affects mechanical qualities like higher strength and hardness [8]. On the other hand, problems like stability or agglomeration may result from the higher surface energy. Achieving the appropriate material characteristics and performance in synthesis requires careful control of these aspects. The ratio of surface area to volume and surface energy have to be studied and controlled to maximize the potential of nanomaterials for use in industries including electronics, medication delivery, and catalysis [9].

The process of synthesis and the final features of the materials are influenced by thermodynamics and kinetics, which are vital concepts in the comprehension of nanomaterial production. By examining variations in free energy, thermodynamics reveals the constancy and phase behaviour of nanomaterials. It aids in the prediction of the most permanent forms and circumstances for the formation and persistence of nanomaterials [10]. The choice of reaction conditions that encourage the production of desirable nanostructures is guided by the principles of thermodynamics. On the other hand, kinetics deals with the synthesis routes and the speeds at which these tiny substances are created. It entails comprehending the reaction processes and energy barriers that govern the rate and efficiency of nanomaterial production. To maintain steady quality and performance, nanomaterials' size, shape, and homogeneity must be carefully controlled through rigorous management of both thermodynamic stability and kinetic pathways. For synthesis procedures to be optimized and required material qualities to be achieved for particular applications, these factors must be balanced [11].

2.3 Synthesis Approaches

The synthesis of nanomaterials by top-down approaches involves separating bulk materials into structures at the nanoscale. These techniques begin with bigger chunks of material that are smaller via physical, chemical, or mechanical processes. Typical methods include laser ablation, which vaporizes a target material with powerful laser pulses and allows the vapour to condense into nanoparticles, ball milling, which employs high-energy impacts to grind materials into fine powders, and etching, which removes particular components of material employing chemical or physical methods to generate nanoscale features. Top-down approaches have the benefit of yielding nanomaterials with precise size and form, but they may have limitations in terms of scalability and defect possibility. They are frequently employed in industries like electronics and nanostructured coatings that need fine structural control, despite these difficulties. These techniques offer a useful strategy for creating materials at the nanoscale with particular functional characteristics [12, 13]. The capacity of mechanical milling to produce a variety of nanostructures, including nanoparticles, nanorods, and nanosheets, renders it desirable. It works especially well for preparing materials that are challenging to synthesize in other ways. Nevertheless, owing to the tremendous energy required, the procedure may generate flaws and structural inhomogeneities. Furthermore, perfect control over the particle size and dispersion is necessary.

Lithography is a top-down synthesis method that produces very precise nanoscale structures. It entails covering a substrate with a photosensitive substance called a resist. After that, the resist is subjected to radiation in the form of light or other waves via a covering or pattern, which modifies its chemical composition in a targeted manner. The barrier is generated after exposure, and depending on the kind of resistance (positive or negative), either the exposed or unexposed portions are washed away. On the substrate, this generates a structured pattern. To create complex nanostructures, further processing stages like etching or deposition are necessary to transfer the pattern onto the underlying material. In the semiconductor industry, lithography is commonly employed for developing electronic circuits and microelectromechanical systems (MEMS). Although it provides excellent pattern integrity and resolution, its associated expenses and scalability can present difficulties, especially in large-scale manufacturing [14].

A flexible bottom-up technique for creating thin films and nanomaterials is the sol–gel process. Through a sequence of chemical events, a liquid solution (sol) is transformed into a solid gel matrix. To form a homogenous sol, precursor chemicals, usually metal alkoxides or salts, are first dissolved in a solvent. After the sol is subjected to hydrolysis and condensation processes, a gel a three-dimensional web of linked nanoparticles is developed. To eliminate any leftover solvents and to densify the material, this gel is then dried and heated. The product is a solid nanomaterial with a carefully managed composition and structure. The sol–gel technique is appropriate for creating coatings, ceramics, and sophisticated materials as it gives exact control over the substance's composition, permeability, and shape. It is beneficial for utilisation in electronics, optics, and catalysis due to its versatility and comparatively low processing temperatures [15].

Bottom-up techniques to generate nanomaterials under regulated temperature and pressure configurations include solvothermal and hydrothermal synthesis. In hydrothermal synthesis, reactions take place in an autoclave a sealed vessel in a solution of water at high temperatures and pressures. A large variety of materials, particularly metal oxides and sulfides, can potentially be synthesized using this technique and have excellent purity and well-defined structures. This approach is also referred to as "solvothermal synthesis," which allows the synthesis of compounds that are neither stable nor soluble in water by substituting organic solvents for water. Solvothermal synthesis may provide a wide range of nanomaterials with customized characteristics by varying the types of solvents used and the circumstances of the process. Exact control over crystallinity, morphology, and particle size is possible with both techniques. In materials science, they are commonly utilized to create catalysts, semiconductor materials, and advanced ceramics [16]. High temperatures and pressures accelerate reactions and encourage the creation of intricate nanostructures.

2.3.1 Green Synthesis Methods

An eco-friendly technique to generate nanomaterials is biological synthesis, which uses organic molecules or their extracts to produce nanomaterials. This method mediates the creation of nanoparticles using bacteria, fungi, plants, or enzymes. In particular, natural biochemical processes in plant extracts may convert metal ions into nanoparticles, and the metabolic processes of fungi and bacteria can collect and stabilize nanomaterials. Numerous benefits of biological synthesis include less of an adverse effect on the environment, less energy consumption, and a decrease in the usage of dangerous chemicals. Furthermore, it frequently produces biodegradable and biocompatible nanomaterials, which are beneficial for utilization in environmental cleanup and medical applications. Due to biological system's ability to direct the production and stabilization of nanoparticles with high precision. In general, biosynthesis is a green chemistry-aligned and promising method for producing sustainable nanomaterials [17].

Utilising microwave radiation to speed up chemical processes in order to produce nanomaterials, microwave-assisted synthesis is a cutting-edge method. Using microwave radiation, the reaction mixture is heated quickly and evenly in this approach, creating a high-energy atmosphere that expedites the synthesis process. Microwave radiation is absorbed by the reactant's dielectric qualities, which results in precise and effective heating. Comparing this to conventional heating techniques yields superior material characteristics, faster reaction times, and increased yield. To synthesize nanoparticles, nanowires, and nanocomposites with precise size and shape, microwave-assisted synthesis is especially useful. In addition, the technique has the benefit of low energy consumption, low by-product production, and gentle material synthesis. Due to its adaptability, it may be used in a variety of fields, including electronics, materials research, and catalysis. The synthesis of nano-materials is more accurate and efficient when done with the help of microwaves [18].

2.4 Characterization Techniques for Nanomaterials

2.4.1 X-Ray Diffraction (XRD)

One of the most important methods for describing nanomaterials and learning in-depth details about their crystalline structure is X-ray diffraction (XRD). X-ray diffraction (XRD) is a technique used to detect the crystal lattice characteristics, phase composition, and structural aspects of a nanomaterial sample by directing X-rays at it and analyzing the diffraction patterns that result. X-rays are dispersed in certain directions when they interact with a crystal's periodic atomic planes, creating a diffraction pattern that may be observed and studied. The resultant pattern provides details on the size of the crystallites, the presence of different phases or impurities, and the unit cell dimensions of the material. Because XRD may shed light on phase transitions, crystal orientation, and strain inside nanoparticles, it is extremely benefi-cial for investigating nanomaterials. Through the analysis of diffraction pattern peak locations and intensities, researchers can get crucial structural information that is essential for comprehending and refining nanomaterial characteristics and uses [19].

2.4.2 Scanning Electron Microscopy (SEM)

High-resolution imaging methods for characterizing nanomaterials are provided by scanning electron microscopy (SEM), a potent imaging tool. SEM operates by moving a concentrated electron beam throughout a sample's surface. Secondary electrons, backscattered electrons, and X-rays are produced when electrons contact with the material; these byproducts are gathered to create intricate pictures. At the nanoscale scale, material topography, morphology, and texture are shown by these photographs. Researchers can examine minute details including particle size, shape, and surface characteristics because of the high-resolution pictures that SEM gives.

Furthermore, by using Energy-Dispersive X-ray spectroscopy (EDS) to analyze elements, SEM may be fitted with a variety of detectors to provide compositional information. Uses in material science, nanotechnology, and quality control need this method since it is so useful for assessing the quality, structure, and dispersion of nanomaterials [20].

2.4.3 Transmission Electron Microscopy (TEM)

An advanced imaging method for examining the internal structure of nanoparticles at the atomic level is transmission electron microscopy or TEM. With a transmission electron microscope (TEM), an ultra-thin sample is exposed to an electron beam, which is then utilized to create finely detailed pictures. The Transmission Electron Microscopy (TEM) shows interior characteristics including flaws, nanostructures, and lattice structures, while the Scanning Electron Microscopy (SEM) only displays surface information. Differences in electron density inside the sample cause contrast in TEM pictures, which makes it possible to see minute structural features and crystallographic information. TEM is crucial for examining the size, shape, and morphology of nanomaterials since it can reach resolutions as low as the atomic level.

For a thorough examination of crystal structure and composition, TEM can be used in conjunction with methods like as energy-dispersive X-ray spectroscopy (EDS) and selected area electron diffraction (SAED). TEM is an essential tool for materials science and nanotechnology research and development [21].

2.4.4 Atomic Force Microscopy (AFM)

AFM, or atomic force microscopy, is a scanning probe method that provides high-resolution surface topography and property characterization of nanomaterials. A pointed probe or tip is scanned across the sample surface for AFM to function. The flexible cantilever on which the tip is placed flexes in reaction to forces acting on the sample surface, including electrostatic or van der Waals forces. By monitoring the cantilever's displacement using a laser beam bounced off the cantilever and into a photodetector, these interactions are found. With nanometre to atomic level resolution, AFM produces three-dimensional pictures of the surface that show characteristics including shape, mechanical characteristics, and roughness on the surface. AFM may also operate in several modes, including force spectroscopy to evaluate local mechanical characteristics and tapping mode for fragile samples [22].

2.4.5 Spectroscopic Techniques (FTIR, UV–Vis, Raman)

Spectroscopic methods that include Raman, UV–Vis, and FTIR are essential for examining the electrical and chemical characteristics of nanomaterials. By analyzing the absorption of infrared light, Fourier Transform Infrared Spectroscopy (FTIR) finds molecular vibrations and functional groups. It assists in distinguishing between the organic and inorganic components of nanomaterials by shedding light on chemical bonding and molecular structure. To identify electronic transitions in nanomaterials, Ultraviolet-Visible Spectroscopy, or UV–Vis, examines the absorption of ultraviolet and visible light. It is often used to characterise nanoparticles and nanostructures and is helpful for researching optical features including bandgap energy and plasmon resonance. To investigate a material's vibrational modes, monochromatic light is scattered during the Raman Spectroscopy process. It offers details on the chemical environment, crystal structure, and molecular makeup, providing insights on the characteristics of the material [23].

2.5 Challenges and Limitations in Nanomaterial Synthesis

A major obstacle to the synthesis of nanomaterials is reproducibility, which affects the uniformity and dependability of the materials that are created. It can be challenging to achieve consistent size, shape, and characteristics among batches since nanoscale processes are sensitive to even small changes in experimental circumstances. Temperature, pressure, precursor concentration, reaction time, and other variables might affect the result and cause inconsistent results. Furthermore, even little modifications to the apparatus or synthesis techniques can produce notable changes in the properties of nanomaterials. Precise control over these factors and a deep comprehension of the synthesis mechanisms are necessary to ensure repeatability. To solve these problems, standardizing procedures and putting strict quality control systems in place are crucial. Consistent findings are still difficult to get, especially when moving from laboratory to industrial production.

In order to confidently employ nanomaterials in sectors such as energy systems, electronics, and medicine, reproducibility issues need to be addressed [24]. When scaled up, many synthesis techniques that perform well at small scales have challenges related to cost, homogeneity, and consistency. For example, methods like chemical vapour deposition or sol–gel procedures can provide high-quality nanomaterials in tiny amounts, but they are not as efficient or repeatable in an industrial setting. Changes to equipment, procedures, and reaction conditions are frequently necessary when scaling up, and these changes can lower yield and add unpredictability. Cost-effectiveness is also an issue since, when manufacturing in big quantities, some technologies may become unaffordable. Improving synthesis methods, creating reliable process control systems, and guaranteeing effective material management and quality are all necessary to address scalability [25].

In order to commercialise nanomaterials and incorporate them into useful applications in sectors like electronics, medicines, and energy, these obstacles must be overcome.

Purity and yield are important aspects of the synthesis of nanomaterials that affect the end product's functionality and applicability. The lack of impurities or undesirable byproducts in the nanomaterial can be referred to as purity. Due to the intricacy of the synthesis procedures and the possibility of contaminants from reacting agents, solvents, or equipment, achieving high purity can be difficult. The characteristics of the material may be impacted by contaminants, which might result in inconsistent or subpar performance in some uses. About the starting ingredients, productivity is the amount of the intended nanomaterial that is produced throughout the synthesis process. Achieving a high yield is crucial for manufacturing to be cost-effective, but it can be challenging owing to things like partial reactions, material loss during processing, or inefficient scaling up. Optimizing synthesis conditions, employing premium precursors, and applying strict purification and separation procedures are necessary to increase both purity and yield. Maintaining the performance and commercial feasibility of nanomaterials in a range of applications requires striking a balance between these aspects [26].

The synthesis of nanomaterials raises important questions about safety and the environment because of the possible effects on ecosystems and human health. The possibility of contaminants from synthesis processes, issues with trash disposal, and the accumulation of nanomaterials in the environment that might upset ecological systems are among the environmental concerns. A combination of their unusual reactivity and durability, nanoparticles might be dangerous for the quality of the soil, water, and air. The handling of dangerous chemicals used in synthesis, the possible toxicity of nanomaterials to people and animals, and the hazards of exposure or inhalation are the main safety issues. Nanomaterials tiny size and strong reactivity might result in unanticipated health risks, such as skin or respiratory problems [27].

2.6 Case Studies

2.6.1 Synthesis of Metal Nanoparticles

These studies shown in Table 2.1 explore different synthesis techniques for metal nanoparticles to enhance space radiation protection. Electrodeposition method was used to synthesize silver nanoparticles, aiming to improve radiation shielding [28]. Laser ablation method for gold nanoparticles synthesis and further used to investigate their protective properties [29]. Titanium dioxide by chemical method for their radiation shielding effectiveness [30]. Palladium nanoparticle synthesized by chemical synthesis and arc deposition contributing to radiation protection research [31].

Table 2.1 Synthesis of metal nanoparticles and their applications

S. No.	Metal nanoparticle	Synthesis method	Application	References
1	Silver nanoparticle	Electrodeposition method	Space device	[28]
2	Gold nanoparticle	Laser ablation	Space radiation protection	[29]
3	Titanium di-oxide	Chemical method	Radiation shielding	[30]
4	Palladium nanoparticle	Chemical synthesis	Radiation protection	[31]
5	Palladium nanoparticle	Arc discharge	Radiation protection	[32]

2.6.2   Synthesis of Carbon Nanotubes

The process for generating carbon nanotubes (CNTs) comprises several important steps, each of which has an impact on the final nanotube's electrical and structural characteristics. The most often used methods are arc discharge, laser ablation, and chemical vapour deposition (CVD). In chemical vapour deposition (CVD), a metal catalyst typically iron, cobalt, or nickel helps a carbon source's gas such as methane or acetylene decompose at high temperatures. Both the length and diameter of the nanotubes may be precisely controlled by this method. By shining a high-energy laser onto a graphite target in a controlled setting, graphite is vaporized and eventually condenses into carbon nanotubes (CNTs). This process is known as laser ablation. With fewer flaws, this process can provide high-quality nanotubes. Arc discharge creates carbon nanotubes (CNTs) by creating an electric arc between two graphite electrodes in an environment of inert gas. Although this method can provide significant quantities of nanotubes, it frequently produces a combination of different forms of carbon nanostructures. Every technique affects the CNTs' length, purity, and alignment, which affects how well-suited they are for different uses in electronics, material science, and nanotechnology [33].

2.6.3   Synthesis of Quantum Dots

The process of creating quantum dots (QDs) requires exact control over the dimensions and morphology of these semiconductor particles at the nanoscale, which influences their optical and electrical characteristics. The most popular method, called colloidal synthesis, involves dissolving precursors in a solution and heating them to create nanoparticles. With this technique, temperature, reaction time, and precursor concentration can all be precisely adjusted to precisely tune QD size. Another method is chemical vapour deposition (CVD), which produces uniform, high-quality dots by reacting gaseous precursors on a substrate to create QDs. High-pressure and high-temperature water solutions are used in hydrothermal synthesis to generate QDs, providing control over their size and composition. Sol–gel techniques, which are often employed to fabricate thin films, use a sol solution that proceeds through gelation to generate QDs. The dimension, shape, and confinement quantum effects of QDs are influenced by each approach, which in turn affects the applications of QDs in biomedical imaging, optoelectronics, and sensing technologies [34].

2.7 Future Trends in Nanomaterial Synthesis

With improvements in accuracy and usefulness, future developments in nanomaterial production have the potential to completely transform the industry. Thanks to advancements in molecular beam epitaxy and self-assembly techniques, bottom-up synthesis methods are becoming more controllable over the production of nanostructures. Chemical vapour deposition (CVD) innovations are making it possible to produce large-scale, high-quality nanomaterials with customised features. Green synthesis techniques, which emphasise ecologically friendly approaches that use biological agents to create nanoparticles with less ecological effect, are becoming more and more popular. 3D printing and additive manufacturing are becoming more and more potent instruments for building intricate nanostructures with exact control over material emplacement. Furthermore, more advanced plasma-based methods are being developed to produce nanoparticles with greater control. The development of novel substances as a consequence of these advances will benefit everything from medicinal devices and ecological restoration to better electronics and storage of energy [35].

The development of substances and chemicals is being revolutionized by the use of AI and machine learning in synthesis processes. Large-scale datasets may be analyzed by AI algorithms and machine learning models to find trends and accurately forecast response times. This capacity improves reaction condition optimization, resulting in more productive and economical operations. Artificial intelligence (AI) powered technologies speed up discovery by making it easier to build novel compounds by anticipating their characteristics and their uses. Moreover, machine learning helps to increase repeatability and decrease human error in synthesis workflow automation. Furthermore, AI models can continually analyze and modify settings in real-time, guaranteeing peak performance and cutting down on waste. These developments not only increase synthesis efficiency but also promote sustainability by lowering the need for labor-intensive trial-and-error testing and preserving resource [36].

Chemical process resource conservation and environmental effect reduction are given top priority in sustainable and environmentally friendly synthesis methods. These techniques place a focus on energy conservation, the use of renewable resources, and a decrease in dangerous products. Green chemistry concepts, including using non-toxic solvents, cutting waste with effective reactions, and recycling byproducts, are important tactics. By improving reaction selectivity and efficiency, catalysis plays a vital role in lowering the requirement for redundant reagents. Furthermore, energy-saving methods such as electrochemical procedures and microwave-assisted synthesis are becoming more popular. Using renewable energy sources, including wind and solar energy, promotes sustainability even more. Greener techniques are also aided by developments in bioengineering and the use of natural catalysts or enzymes. All in all, these strategies seek to develop more environmentally friendly industrial processes while reducing adverse impacts on the environment and encouraging the usage of safer, more efficient technologies [37].

2.8 Conclusion

Resource utilization and environmental impact reduction are the main goals of sustainable and environmentally friendly synthesis techniques. The utilization of green chemistry concepts, which prioritize the use of non-toxic materials, decreased waste creation, and energy conservation, is one of the main aspects. Reaction efficiency may be increased and energy requirements can be decreased with the use of catalysis and renewable energy sources. Energy-saving and environmentally friendly techniques include electrochemical processes and microwave-assisted synthesis. Furthermore, by using renewable biological assets, developments in bioengineering and the application of natural catalysts promote sustainability even further. All things considered, the goal of these tactics is to promote safer industrial processes, lessen the production of dangerous byproducts, and aid in environmental protection all of which will eventually lead to a more sustainable future for chemical synthesis.

Technology breakthroughs and a growing emphasis on sustainability will drive major advancements in the synthesis of nanomaterials in the future. Innovative techniques like chemical vapour deposition and self-assembly, which are based on bottom-up approaches, provide fine control over the structures of nanomaterials, allowing for the production of highly customized and useful materials. Green chemistry developments are driving the creation of environmentally friendly synthesis techniques, lowering dependency on hazardous chemicals, and minimizing negative effects on the environment. Artificial intelligence and machine learning together are improving synthesis processes and improving prediction skills. Furthermore, new applications in domains like medicine, electronics, and energy storage are promised by the research of innovative nanomaterials like 2D materials and hybrid systems. Enhancements in scalability and cross-disciplinary research will be essential for turning these breakthroughs into real-world, large-scale applications, expanding the possibilities of nanomaterials to address intricate global issues.

References

1. Al-Douri Y (2022) Nanomaterials, vol 13. Springer Nature, Singapore, pp 10–1007
2. Tejashwini DM, Harini HV, Nagaswarupa HP, Naik R, Deshmukh VV, Basavaraju N (2023) An in-depth exploration of eco-friendly synthesis methods for metal oxide nanoparticles and their role in photocatalysis for industrial dye degradation. Chem Phys Impact 7:100355
3. Vázquez-López A, García-Carrión M, Hall E, Yaseen A, Kalafat I, Taeño M, Zhu J, Zhang X, Arici E, Taskin OS, Maestre D, Nogales E, Hidalgo P, Ramírez-Castellanos J, Méndez B, Yuca N, Karazhanov S, Marstein ES, Cremades A (2021) Hybrid materials and nanoparticles for hybrid silicon solar cells and Li-ion batteries. J Energy Power Technol 3(2):1–25
4. Pokrajac L, Abbas A, Chrzanowski W, Dias GM, Eggleton BJ, Maguire S, Maine E, Malloy T, Nathwani J, Nazar L, Sips A, Sone J, van den Berg A, Weiss PS, Mitra S (2021) Nanotechnology for a sustainable future: addressing global challenges with the international network4sustainable nanotechnology
5. Hull M, Bowman D (eds) (2014) Nanotechnology environmental health and safety: risks, regulation and management. William Andrew
6. Bera D, Qian L, Tseng TK, Holloway PH (2010) Quantum dots and their multimodal applications: a review. Materials 3(4):2260–2345
7. Terna AD, Elemike EE, Mbonu JI, Osafile OE, Ezeani RO (2021) The future of semiconductors nanoparticles: synthesis, properties and applications. Mater Sci Eng B 272:115363
8. Liu CJ, Burghaus U, Besenbacher F, Wang ZL (2010) Preparation and characterization of nanomaterials for sustainable energy production
9. Satpute HT, Gadgile DP, Arsule AD, Dabhade VF, Urkude GP, Poul SV, Jadhav NS, Karhale GA, Pawar PV, Zangade SW, Nirwal AU (2023) Emerging trends in basic sciences
10. Tejashwini DM, Harini HV, Nagaswarupa HP, Naik R (2024) Hybrid calcium magnesium ferrite: a multifaceted nanomaterial for energy, sensing, and environmental applications. Results Chem 7:101456
11. Matyjaszewski K (2005) Macromolecular engineering: from rational design through precise macromolecular synthesis and processing to targeted macroscopic material properties. Prog Polym Sci 30(8–9):858–875
12. Biswas A, Bayer IS, Biris AS, Wang T, Dervishi E, Faupel F (2012) Advances in top–down and bottom–up surface nanofabrication: techniques, applications & future prospects. Adv Colloid Interface Sci 170(1–2):2–27
13. Yadav TP, Yadav RM, Singh DP (2012) Mechanical milling: a top down approach for the synthesis of nanomaterials and nanocomposites. Nanosci Nanotechnol 2(3):22–48
14. Baek D, Lee SH, Jun BH, Lee SH (2021) Lithography technology for micro- and nanofabrication. In: Nanotechnology for bioapplications, pp 217–233
15. Raghavendra N, Nagaswarupa HP, Shekhar TS, Mylarappa M, Surendra BS, Prashantha SC, Ravikumar CR, Kumar MRA, Basavaraju N (2021) Development of clay ferrite nanocomposite: electrochemical, sensors and photocatalytic studies. Appl Surf Sci Adv 5:100103
16. Ndlwana L, Raleie N, Dimpe KM, Ogutu HF, Oseghe EO, Motsa MM, Mamba BB (2021) Sustainable hydrothermal and solvothermal synthesis of advanced carbon materials in multidimensional applications: a review. Materials 14(17):5094
17. Tejashwini DM, Harini HV, Nagaswarupa HP, Naik R, Chidananda B (2024) A comparative study of green and chemical approaches for photocatalytic activity of novel hybrid bismuth magnesium ferrites (BiMgFe$_2$O$_4$) nanoparticles for Acid Red-88 dye degradation. Results Chem 7:101267
18. Zhu YJ, Chen F (2014) Microwave-assisted preparation of inorganic nanostructures in liquid phase. Chem Rev 114(12):6462–6555
19. Deshmukh VV, Tejashwini DM, Nagaswarupa HP, Naik R, Al-Kahtani AA, Kumar YA (2024) Sr and Fe substituted LaCoO$_3$ nano perovskites: electrochemical energy storage and sensing applications. J Energy Storage 89:111724

20. Tejashwini DM, Nagaswarupa HP, Naik R, Basavaraju N, Alodhayb AN, Pandiaraj S, Goud BS, Kim JH (2024) Synthesis, characterization, wastewater treatment & plant growth application of $ZnFe_2O_4/Bi_2O_3$ nanocomposite. Desalination Water Treat 320:100681
21. Biswas K, Sivakumar S, Gurao N (eds) (2022) Electron microscopy in science and engineering. Springer
22. Sulania I, Yadav RP, Karn RK (2018) Atomic and magnetic force studies of Co thin films and nanoparticles: understanding the surface correlation using fractal studies. In: Handbook of materials characterization, pp 263–291
23. Saif FA, Yaseen SA, Alameen AS, Mane SB, Undre PB (2021) Identification and characterization of Aspergillus species of fruit rot fungi using microscopy, FT-IR, Raman and UV–Vis spectroscopy. Spectrochim Acta Part A Mol Biomol Spectrosc 246:119010
24. Gao X, Lowry GV (2018) Progress towards standardized and validated characterizations for measuring physicochemical properties of manufactured nanomaterials relevant to nano health and safety risks. NanoImpact 9:14–30
25. Saldanha PL, Lesnyak V, Manna L (2017) Large scale syntheses of colloidal nanomaterials. Nano Today 12:46–63
26. García-Quintero A, Palencia M (2021) A critical analysis of environmental sustainability metrics applied to green synthesis of nanomaterials and the assessment of environmental risks associated with the nanotechnology. Sci Total Environ 793:148524
27. Ahmed SF, Mofijur M, Rafa N, Chowdhury AT, Chowdhury S, Nahrin M, Islam ABMS, Ong HC (2022) Green approaches in synthesising nanomaterials for environmental nanobioremediation: technological advancements, applications, benefits and challenges. Environ Res 204:111967
28. Marciano FR, Bonetti LF, Pessoa RS, Marcuzzo JS, Massi M, Santos LV, Trava-Airoldi VJ (2008) The improvement of DLC film lifetime using silver nanoparticles for use on space devices. Diam Relat Mater 17(7–10):1674–1679
29. Yilmaz AH, Ortaç B, Mutlu S, Yilmaz SS (2023) Synthesis of polyethylene-based materials, ion exchanger, superabsorbent, radiation shielding, and laser ablation applications. In: Polyethylene-new developments and applications. IntechOpen
30. More CV, Botewad SN, Akman F, Agar O, Pawar PP (2023) UPR/titanium dioxide nanocomposite: preparation, characterization and application in photon/neutron shielding. Appl Radiat Isot 194:110688
31. Zhang Z, Cui X, Yuan W, Yang Q, Liu H, Xu H, Jiang HL (2018) Encapsulating surface-clean metal nanoparticles inside metal–organic frameworks for enhanced catalysis using a novel γ-ray radiation approach. Inorg Chem Front 5(1):29–38
32. Rojas Marin JV (2011) Production and characterization of supported palladium nanoparticles on multiwalled carbon nanotubes by gamma irradiation
33. Shah KA, Tali BA (2016) Synthesis of carbon nanotubes by catalytic chemical vapour deposition: a review on carbon sources, catalysts and substrates. Mater Sci Semicond Process 41:67–82
34. Nagpal R, Gusain M (2022) Synthesis methods of quantum dots. In: Graphene, nanotubes and quantum dots-based nanotechnology. Woodhead Publishing, pp 599–630
35. Baig N, Kammakakam I, Falath W (2021) Nanomaterials: a review of synthesis methods, properties, recent progress, and challenges. Mater Adv 2(6):1821–1871
36. Mann V, Sales-Cruz M, Gani R, Venkatasubramanian V (2024) ESFILES: intelligent process flowsheet synthesis using process knowledge, symbolic AI, and machine learning. Comput Chem Eng 181:108505
37. Khatami M, Iravani S (2021) Green and eco-friendly synthesis of nanophotocatalysts: an overview. Comments Inorg Chem 41(3):133–187

Chapter 3
Characterization of Metal Oxide Nanomaterials

Abstract With a focus on their applicability to thermoluminescence (TL) applications, this work examines the structural, morphological, and compositional characteristics of several metal oxide nanomaterials. X-ray diffraction (XRD) was used for structural analysis. The results showed different crystalline states and average grain sizes, two noteworthy discoveries were that ZrO_2 had a crystalline size of 11.62 nm and NiO had considerable particle growth to 35.9 nm. Phenotypic investigation using Fourier transform infrared spectroscopy (FTIR) and scanning electron microscopy (SEM) revealed information on surface functional groups and particle agglomeration. The presence of particular metal–oxygen interactions and hydroxyl groups on the surface was confirmed by FTIR spectra, while SEM pictures revealed diverse particle morphologies and agglomeration tendencies. These characterizations have an important bearing on how well TL materials operate as well as how sensitive they are for radiation monitoring.

3.1 Introduction

Methods for characterization of structure, morphology, and composition are essential for improving thermoluminescence (TL) materials since they have a direct impact on radiation identification efficiency and dependability. Scanning electron microscopy (SEM) and X-ray diffraction (XRD) are two structural analysis techniques that shed light on the crystal lattice, phase purity, and particle form of TL materials. To understand whether radiation-induced carriers of charge are held and released, for example, XRD is useful in detecting phase transitions and crystalline formations. SEM, on the other hand, provides information on the grain size and surface morphology, which have an impact on the material's internal energy transfer and light emission processes. With the use of such precise structural data, TL materials may be customized to function at their best by guaranteeing homogeneity and minimizing flaws that could otherwise lower luminescence effectiveness [1, 2].

The assessment of the influence of the material's physical shape on its TL characteristics is also influenced by morphological characterization. High-resolution photographs of nanostructures and surface topographies are made possible by methods like transmission electron microscopy (TEM) and atomic force microscopy (AFM), which help researchers comprehend how ionized radiation interacts with materials. The chemical composition and doping levels are checked using compositional analysis, which includes energy-dispersive X-ray spectroscopy (EDX) and X-ray photoelectron spectroscopy (XPS), to guarantee the proper luminous performance. Precise chemical information aids in maximizing doping concentrations and understanding various components that affect the TL response. To develop and improve TL materials and increase their sensitivity, accuracy, and overall efficacy in detecting radiation applications, these characterization approaches work together like magic [3].

3.2 Structural Characterisations of Metal Oxide Nanomaterials

3.2.1 Powder X-Ray Diffraction (XRD) Analysis of Metal Oxide Nanoparticles

PXRD Patterns

ZrO$_2$

The XRD pattern of nanoparticles of ZrO$_2$ sample is given in Fig. 3.1. XRD has been used to characterize the structure of ZrO$_2$ nanoparticles. The analysis of the spectrum clearly demonstrates that, the synthesized ZrO$_2$ nanoparticle is identical and could be indexed to the standard ZrO$_2$ with monoclinic structure (JCPDS No.: 37-1484). The prominent peaks have been utilized to estimate the grain size of sample with the help of Scherrer equation. $D = K\lambda/(\beta \cos \theta)$ where, K is constant (0.9), λ is the wavelength ($\lambda = 1.5418$ Å) (Cu Kα), β is the full width at the half maximum of the line ($\beta = 99.2180$), θ is the diffraction angle ($\theta = 30$). The grain size was found to be 11.62 nm, Sharpness of XRD peaks indicates that particles are having in crystalline nature [4].

CrAl$_2$O$_4$

Figure 3.2 shows the XRD diagrams of the compound calcinated at 350, 500, 700 and 800 °C. It can be seen that the phase formation is complete at 700 °C and further increase in calcination temperature does not improve the crystallinity of the compound. All peaks are well defined pointing to the high crystalline nature of the sample. The XRD pattern was compared with JCPDS files (file No. 74-1138) and all the diffraction peaks are in good match with the reported result and the peaks were indexed with the help of these reported patterns. Rietveld refinement was carried out to verify the formation of spinel phase and also to determine the cation distribution or displacement of chromate ion. The lattice parameter obtained from the Rietveld

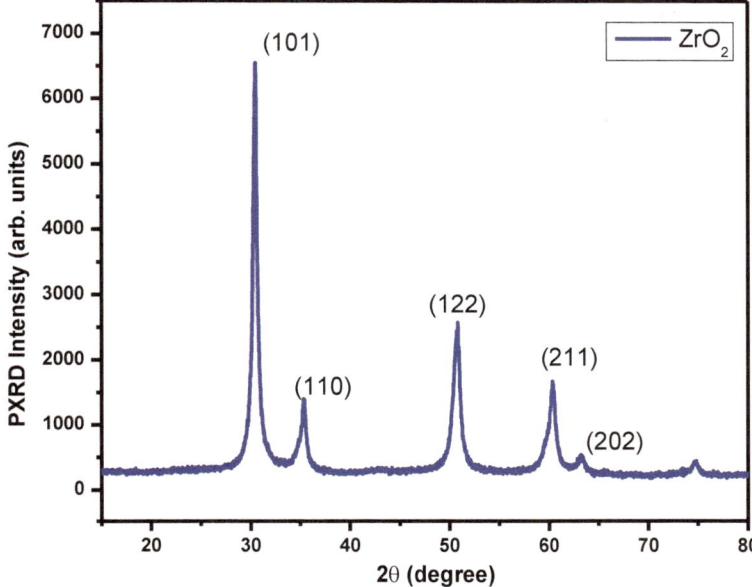

Fig. 3.1 XRD pattern of nanoparticles of ZrO_2 sample

analysis is 8·098 Å. The reported value of lattice parameter is 8·099 Å according to the JCPDS data and this close agreement indicates the phase purity of the chromium aluminate [5].

$ZrAl_2O_4$

Figure 3.3 shows the XRD diagrams of the compound calcinated at 350, 500, 700 and 800 °C. It can be seen that the phase formation is complete at 700 °C and further increase in calcination temperature does not improve the crystallinity of the compound. All peaks are well defined pointing to the high crystalline nature of the sample. The XRD pattern was compared with JCPDS files and all the diffraction peaks are in good match with the reported result and the peaks were indexed with the help of these reported patterns. Rietveld refinement was carried out to verify the formation of spinel phase and also to determine the cation distribution or displacement of zirconium ion. The lattice parameter obtained from the Rietveld analysis is 8·098 Å. The reported value of lattice parameter is 8·099 Å according to the JCPDS data and this close agreement indicates the phase purity of the zirconium aluminate. The average crystal size is 11.6 nm [6].

AgZnO

PXRD patterns of ZNPs shown in Fig. 3.4. It was evident that no impurity peaks/no others phases were observed, indicating the formation of high purity ZnO product. Further, the diffraction peaks appear at ~ 2θ = 31.8, 34.41, 36.27, 47.62, 56.63, 62.93, 66.51 and 69.22 corresponding to (1 0 0), (0 0 2), (1 0 1), (1 0 2), (1 1 0),

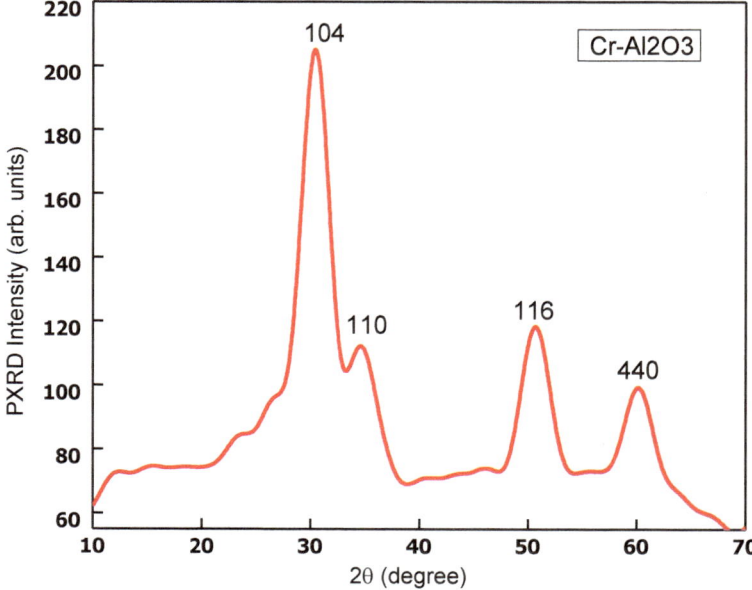

Fig. 3.2 XRD pattern of nanoparticles of $CrAl_2O_4$ sample

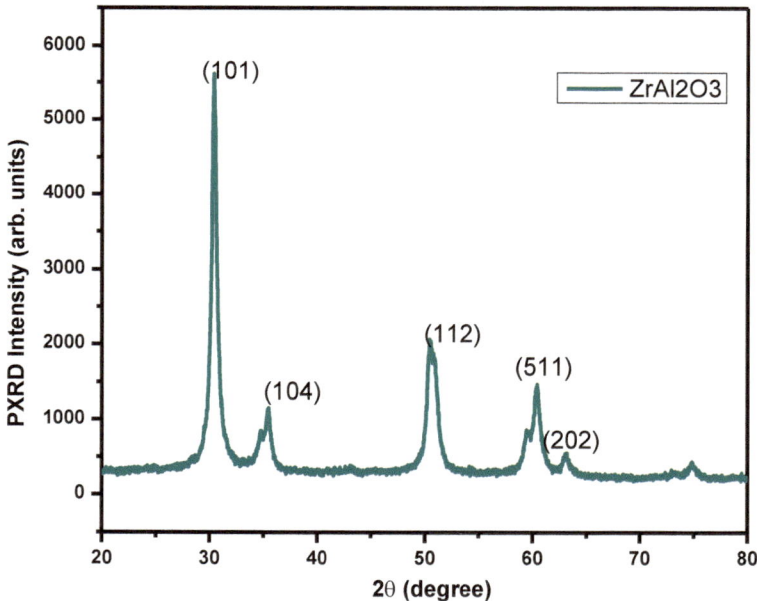

Fig. 3.3 XRD pattern of nanoparticles of $ZrAl_2O_4$ sample

(1 0 3), (2 0 0) and (2 0 1) respectively, were indexed to hexagonal phase, Wurtzite structure (JCPDS card No. 89-1397 with space group P6$_3$mc, No. 186). The average crystallite size (D) was estimated from the line broadening in X-ray powder using Scherrer's formula.

$$D = \frac{K\lambda}{\beta \cos \theta} \tag{3.1}$$

where, 'K'; constant, 'λ'; wavelength of X-rays and 'β'; full width at half maximum (FWHM). The lattice parameters for hexagonal ZnO nanoparticles were estimated from the relation [7].

Ca–ZnO

PXRD patterns of ZNPs shown in Fig. 3.5. It was evident that no impurity peaks/no others phases were observed, indicating the formation of high purity ZnO product. Further, the diffraction peaks appear at ~ 2θ = 31.8, 34.41, 36.27, 47.62, 56.63, 62.93, 66.51, 57.3, 63.0 and 69.22 corresponding to (1 0 0), (0 0 2), (1 0 1), (1 0 2), (1 1 0), (1 0 3), (2 0 0), (220), (311) and (2 0 1) respectively, were indexed to hexagonal phase, Wurtzite structure (JCPDS card No. 89-1397 with space group P6$_3$mc, No. 186). The average crystallite size (D) was estimated from the line broadening in X-ray powder using Scherrer's formula [8, 9].

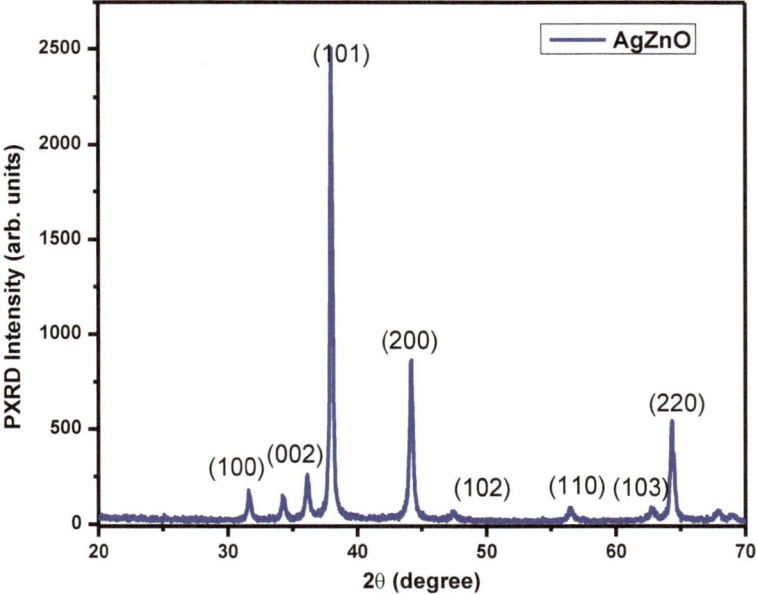

Fig. 3.4 XRD pattern of nanoparticles of AgZnO sample

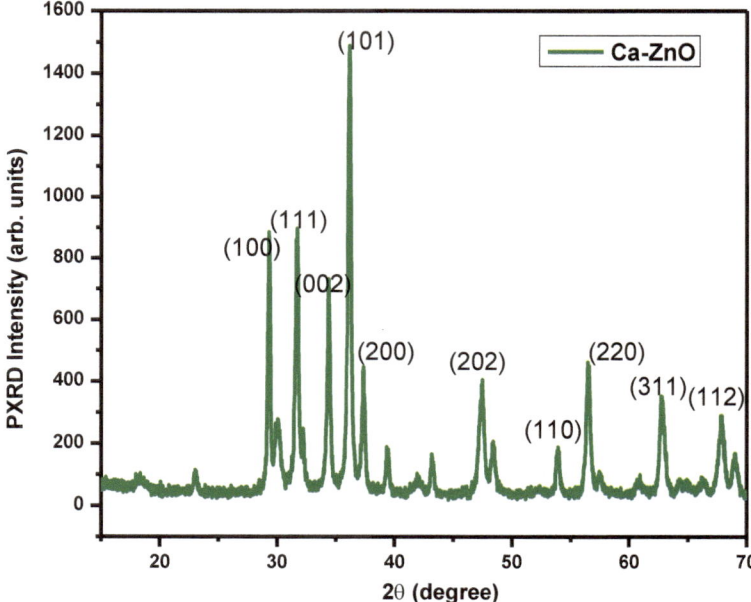

Fig. 3.5 XRD pattern of nanoparticles of Ca–ZnO sample

Zn–MgO (Sol–Gel)

The structure possesses the cubic may be attributed to the different preparation method which may yield different structural defects. The crystalline size was determined from full width of half maximum (FWHM) of the most intense peak obtained by scanning X-ray diffraction pattern. The grain size was calculated by using following Scherrer's formula. $d = 0.9\lambda/\beta \cos\theta$ where, d is the crystalline size, λ is the X-ray wavelength of the Cu Kα source ($\lambda = 1.54056$ Å), β is the FWHM of the most predominant peak at 100% intensity, θ is the Braggs angle at which peak is recorded. The grain size was found to be 14.58 nm [10] (Fig. 3.6).

Ce–MgO

The structure possesses the cubic may be attributed to the different preparation method which may yield different structural defects as illustrated in Fig. 3.7. The crystalline size was determined from full width of half maximum (FWHM) of the most intense peak obtained by scanning X-ray diffraction pattern. The grain size was calculated by using following Scherrer's formula. $d = 0.9\lambda/\beta \cos\theta$ where, d is the crystalline size, λ is the X-ray wavelength of the Cu Kα source ($\lambda = 1.54056$ Å), β is the FWHM of the most predominant peak at 100% intensity, θ is the Braggs angle at which peak is recorded. The grain size was found to be 24.89 nm [11].

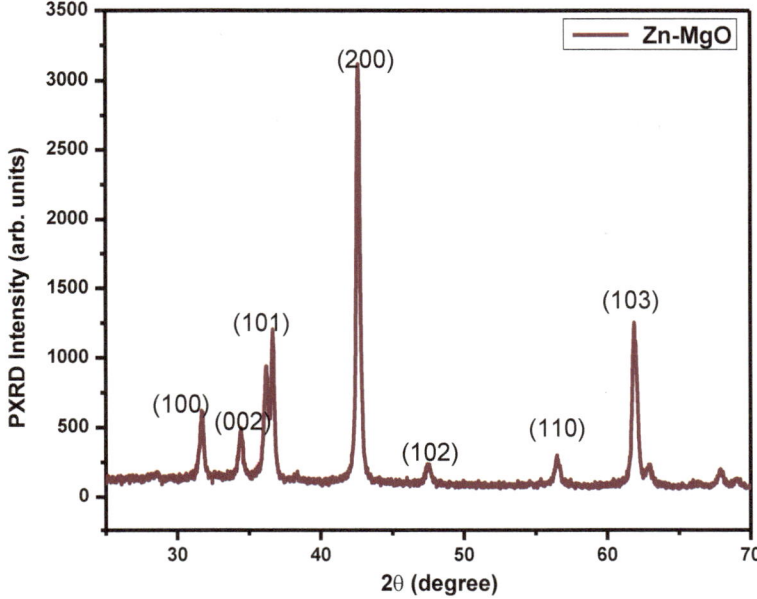

Fig. 3.6 XRD pattern of nanoparticles of Zn–MgO (sol–gel) sample

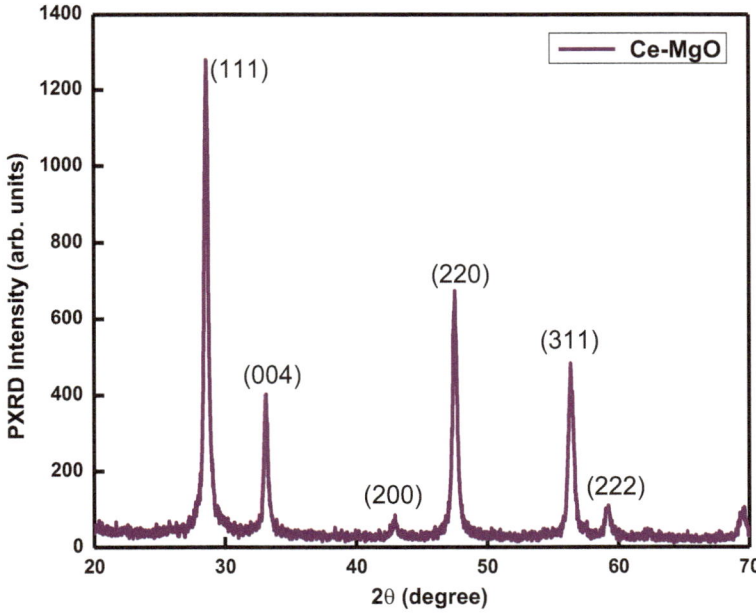

Fig. 3.7 XRD pattern of nanoparticles of Ce–MgO sample

NiO

Figure 3.8 shows the XRD patterns of the precursor NiO nanoparticles product after calcination. The XRD patterns of the calcined sample exhibited sharpened reflection peaks which indicate that a growth in the crystallite sizes of NiO has occurred. The peaks positions appearing at 2θ 37.101, 43.301, 62.871, 76.501, and 79.221 can be readily indexed as (111), (200), (220), (311) crystal planes of the bulk NiO, respectively. All the reflections can be indexed to face-centered cubic (fcc) NiO phase with lattice constant (a): 4.175 Å (space group Fm_3hm [225]) which agrees well with the standard data (JCPDS card No. 47-1049). The sharpness and the intensity of the peaks indicate the well crystalline nature of the prepared sample [12] (Table 3.1).

Fourier Transform Infrared Spectra (FTIR) Analysis of Metal Oxide Nanoparticles

Ca–ZnO

The composition and quality of the material was investigated by the FTIR spectroscopy. The FTIR spectra of pure and cobalt doped ZnO nanoparticles are shown in Fig. 3.9. The spectra show the Zn–O bond appear at around 450–490 cm^{-1}. The band at around 490 cm^{-1} may be related with oxygen vacancy or oxygen deficiency in ZnO. This oxygen deficiency should translate into an enhanced green emission in UV absorption spectra. The peaks at 1350 cm^{-1} correspond to the C–O absorption of ZnO surface. The weak absorption band at 2367 cm^{-1} stands for carbonate that

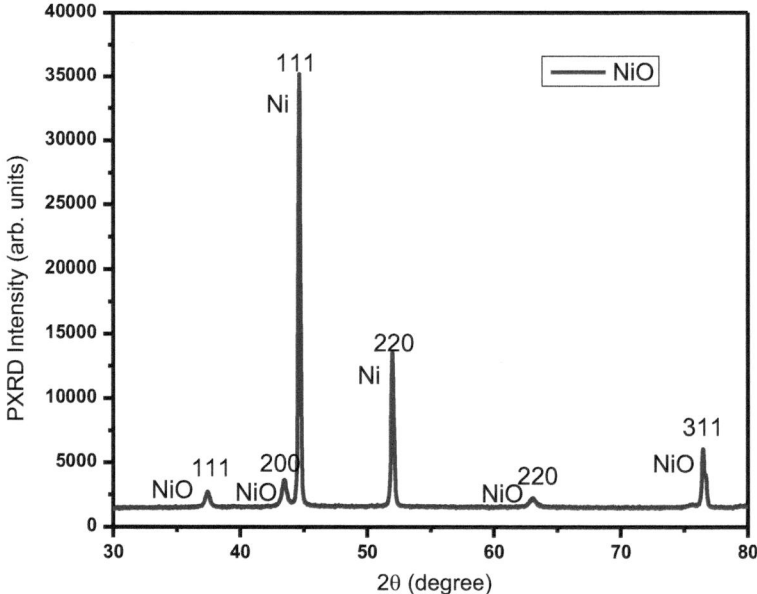

Fig. 3.8 XRD pattern of nanoparticles of NiO sample

Table 3.1 The estimated crystallite size for the above discussed materials

S. No.	Name of the nanoparticles	Size (nm)
1	ZrO_2	11.62
2	$CrAl_2O_4$	27
3	$ZrAl_2O_4$	11.6
4	Bi_2O_3	27.5
5	AgZnO	27.5
6	CaZnO	29.3
7	CeMgO	24.89
8	ZnMgO	5.39
9	ZnMgO (sol–gel)	14.58
10	NiO	35.9

probably comes from the atmospheric carbon dioxide during synthesis. The broad peaks around 3429 cm^{-1} is assigned to the O–H stretching mode of hydroxyl groups 5, 6 and 1634 cm^{-1} (bending) are due to asymmetrical stretching of the zinc carboxylate. In cobalt doping ZnO, the entire peak transmittance % got quenched. Together these results identified the impurities which exist near ZnO surface. No other peak was not observed in the spectra confirms that final products is ZnO nanoparticles [8].

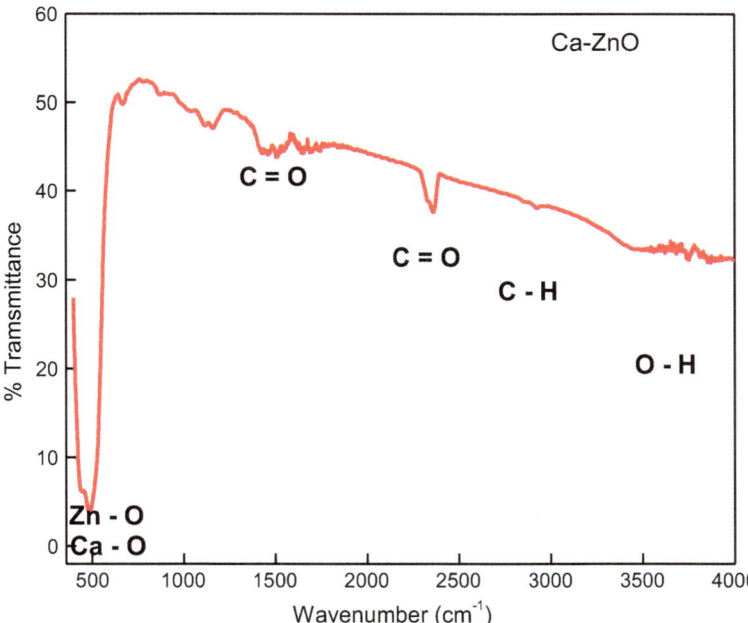

Fig. 3.9 FT-IR pattern of nanoparticles of Ca–ZnO sample

Bi–ZnO

The FT-IR spectrum of the purified Bi–ZnO hybrid nanoparticles with that of the pure material used in the synthesis, the pure material expresses one strong characteristic band at the position of 1109.18 cm^{-1} due to the C–O–C stretching vibration of the ether bonding which usually lies in the range of 1250–1000 cm^{-1} and one sharp characteristic band due to the C–H bending vibration at the position of 1462.65 cm^{-1}. The shape changes and blue shifting in the C–O–C stretching and C–H bending modes [13] may be due to the interactive coordination of the oxygen atoms in the sample as shown in Fig. 3.10.

Ag–ZnO

Synthesized Ag–ZnO nanocomposite material was analysed by FT-IR in the range from 400 to 4000 cm^{-1} at room temperature as shown in Fig. 3.11. The FT-IR spectrum of the Ag: ZnO nanocomposite material contains several bands with remarkable features. The bands between 400 and 750 cm^{-1} correlated to metal oxide bond (ZnO). Bands around 900–1500 cm^{-1} are due to the oxygen stretching and bending frequency. The spectral band at 463.01 cm^{-1} and band at 723.96 cm^{-1} clearly shows the presence of ZnO and Ag ions approximately equal with the reported literature. On doping, stronger and wider absorption, bands observed in the region ~ 723.96 cm^{-1} due to the organic capping of silver. Bands at 1394.82 cm^{-1} and 1507.07 cm^{-1} corresponds to C=O and O–H bending vibrations respectively. The weak band near

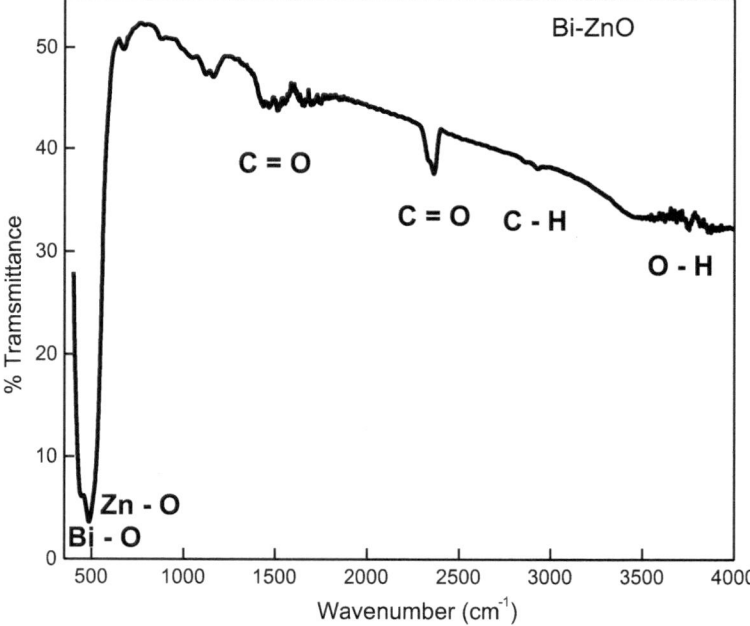

Fig. 3.10 FT-IR pattern of nanoparticles of Bi–ZnO sample

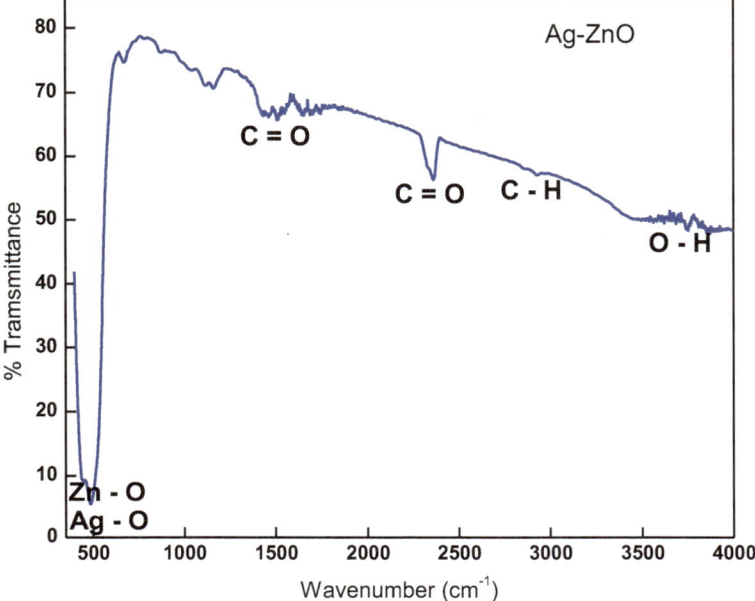

Fig. 3.11 FT-IR pattern of nanoparticles of Ag–ZnO sample

1604.07 cm^{-1} is assigned to H–O–H bending vibration mode were presented due to the adsorption of moisture. Bands at 3465.60 cm^{-1} indicate the presence of OH group. It is evident from the FTIR data that the Zn–O vibrational mode more prominently observed and this clearly concludes a strong doping between Ag and ZnO nanocomposite material [14].

Zn–MgO

FTIR spectra of the synthesized Zn–MgO nanoparticles in the range of 4000–400 cm^{-1} as shown in Fig. 3.12. The broad band at 3450 cm^{-1} is the stretching vibration of O–H group. The peak at 1490 cm^{-1} is due to the O–H bending of water. The peak at 460 cm^{-1} is attributed to the Zn–O–Mg stretching of vibration. The weak band near 1650.07 cm^{-1} is assigned to H–O–H bending vibration mode were presented due to the adsorption of moisture [9].

ZrO$_2$

The typical FTIR spectra in Fig. 3.13 are the finger print of the material. The broad peak exhibited in the FTIR Spectra in the range 3000–3800 cm^{-1} corresponds to the stretching vibration of physically adsorbed –OH with the Zr^{4+} ion on the as shown in Fig. 3.14. The peak centered around 2340–2360 cm^{-1} is attributed to the coupling effect of stretching and bending vibration of –OH groups. The peak at 1690 cm^{-1}

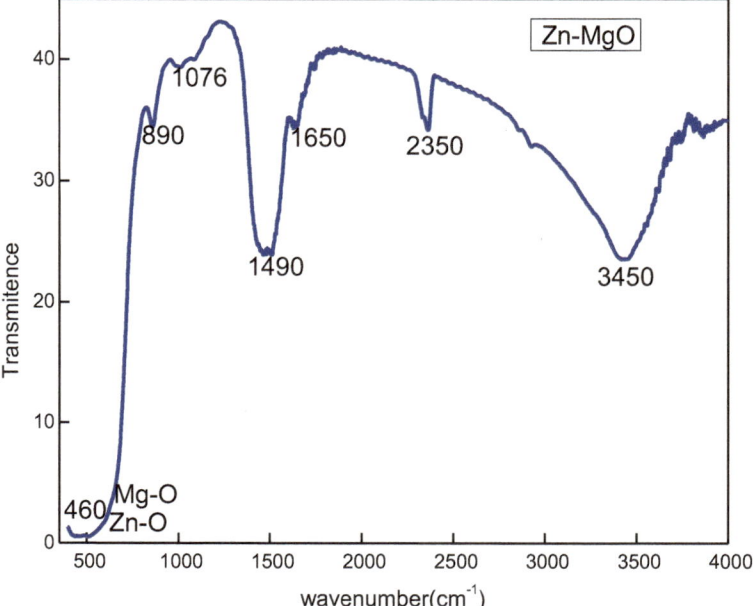

Fig. 3.12 FT-IR pattern of nanoparticles of Zn–MgO sample

is the characteristic peak of H–O–H bending vibration. The peak centered around $1340\,cm^{-1}$ in pure Zirconia alone is due to O–H vibration. The peak around $870\,cm^{-1}$ is the characteristic peak of Zr–O valence vibration. The small absorption peak in the range 500–$451\,cm^{-1}$ is due to Zr–O–Zr vibrations [11, 15].

Bi_2O_3

The IR spectra of the dried calcined powder of bismuth oxide are shown in Fig. 3.14. After drying at 120 °C, the spectrum is complex due to the existence of lots of organic compounds. Band at $3420\,cm^{-1}$ is a characteristic group frequency from the stretch vibration of O–H. The broad band at $2800 \sim 3200\,cm^{-1}$ comes from C–H stretch vibration and the stretch –CH_2 of located at $2930\,cm^{-1}$. The peak of $1390\,cm^{-1}$ is the characteristic ones of NO^{3-} group. The broad one around $825 \sim 465\,cm^{-1}$ originates from the metal–oxygen (Bi–O) vibration [16].

NiO

As shown in figure, the absorption bonds at $430\,cm^{-1}$ are associated to Ni–O vibration bond, but absorption bond at $870\,cm^{-1}$ is assigned to Ni–O–H stretching bond as shown in Fig. 3.15. The above information confirmed formation of pure NiO nanoparticles. Band at $3425\,cm^{-1}$ is a characteristic group frequency from the stretch vibration of O–H. The broad band comes from C–H stretch vibration and the stretch –CH_2 of located at $2360\,cm^{-1}$. The peak centered around $1490\,cm^{-1}$ in pure NiO alone is due to O–H vibration [12].

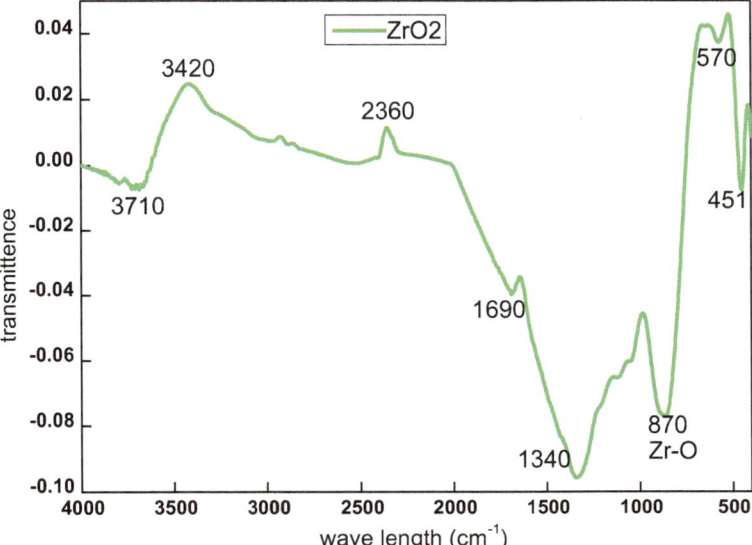

Fig. 3.13   FT-IR pattern of nanoparticles of ZrO$_2$ sample

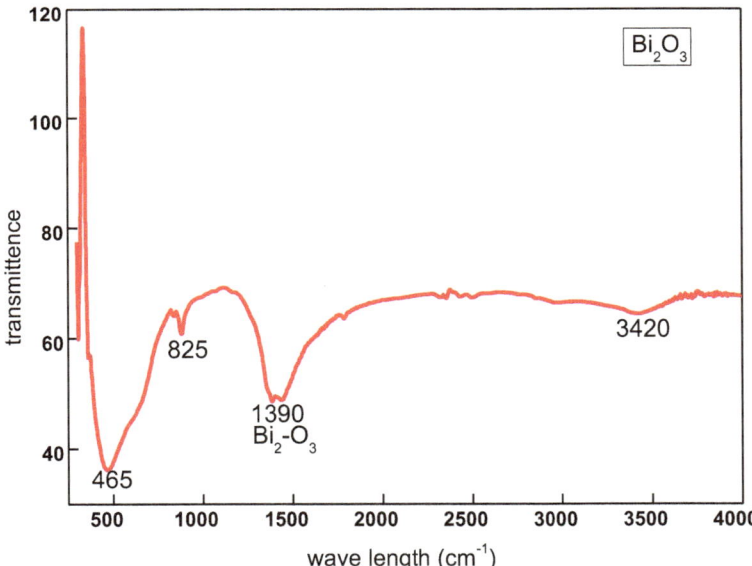

Fig. 3.14   FT-IR pattern of nanoparticles of Bi$_2$O$_3$ sample

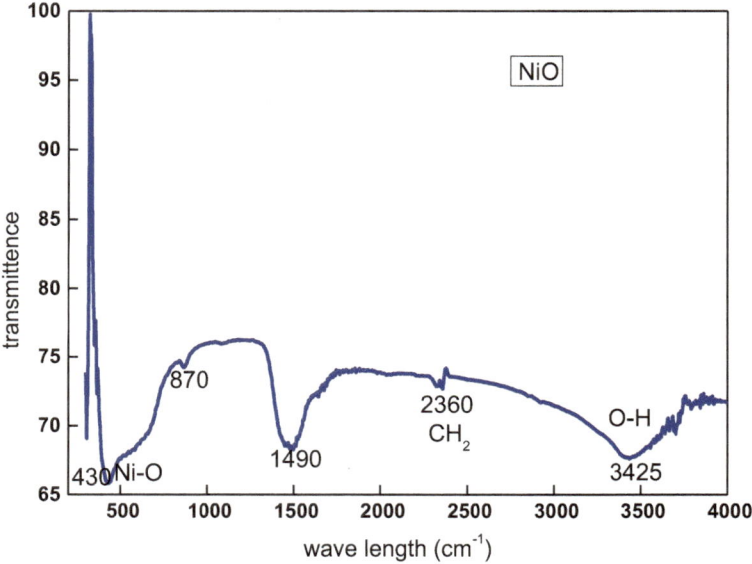

Fig. 3.15   FT-IR pattern of nanoparticles of NiO sample

Zr–Al$_2$O$_4$

The FT-IR spectra show a series of absorption peaks in the range of 400–4000 cm^{-1} as shown in Fig. 3.16. According to the specific frequencies of the absorption peaks, the functional groups existing in the samples can be deduced. Peaks at 1490, 660, and 441 cm^{-1} are present in all samples, and are assigned to the H–O–H bending vibration of adsorbed water, Al–O symmetric stretching vibration $(\nu_1)^1$, Al–O symmetric bending vibration, and Al–O asymmetric stretching vibration, respectively. For the peaks located at 1490 cm^{-1} are attributed to the S=O asymmetric stretching vibration and the S–O symmetric stretching vibration. Band at 3610 cm^{-1} is a characteristic group frequency from the stretch vibration of O–H. The broad band comes from C–H stretch vibration and the stretch –CH$_2$ of located at 2360 cm^{-1}.

Scanning Electron Microscope (SEM) Analysis of Metal Oxide Nanoparticles

ZrO$_2$

The surface morphology of ZrO$_2$ NPs exhibit almost non-uniform porous agglomeration of the particles goes on increases as shown in Fig. 3.17. Which satisfactorily agrees with crystallite size calculated from PXRD values, the agglomeration of particles was generally depends on reaction rate, impurities, charges on the particles etc.

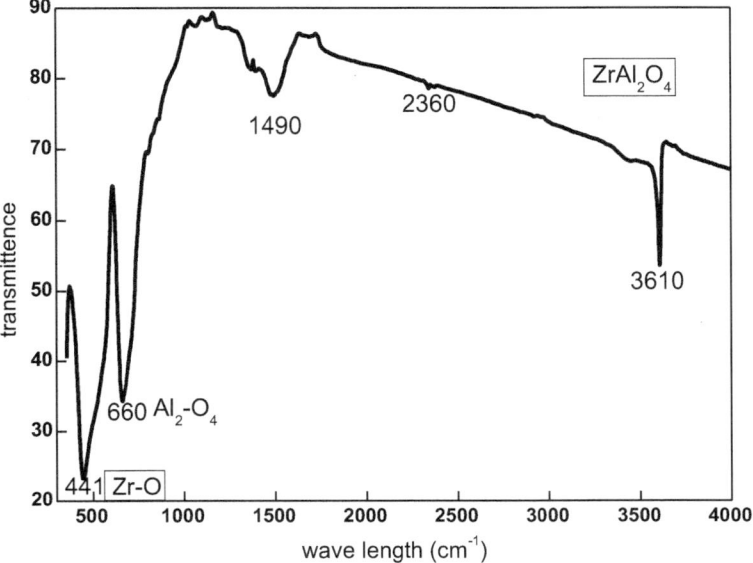

Fig. 3.16 FT-IR pattern of nanoparticles of Zr–Al$_2$O$_4$ sample

Separation of particles depends not only on the particle charge, the charge repulsion between the particles were disabled which leads to coagulation and results in particle agglomeration [17].

CrAl$_2$O$_4$

SEM image of CrAl$_2$O$_4$ NPs exhibit almost porous agglomeration of the particles as shown in Fig. 3.18. Which satisfactorily agrees with crystallite size calculated from

Fig. 3.17 SEM image of nanoparticles of ZrO$_2$ sample

Fig. 3.18 SEM image of nanoparticles of CrAl$_2$O$_4$ sample

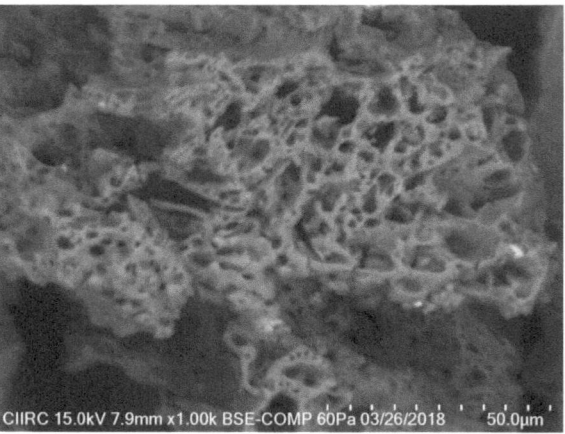

PXRD values, the agglomeration of particles was generally depends on reaction rate, impurities, charges on the particles etc. [18].

ZrAl$_2$O$_3$

SEM image of ZrAl$_2$O$_4$ NPs exhibit almost porous agglomeration of the particles as shown in Fig. 3.19. Which satisfactorily agrees with crystallite size calculated from PXRD values, the agglomeration of particles was generally depends on reaction rate, impurities, charges on the particles etc. [19].

Bi$_2$O$_3$

Bi$_2$O$_3$ exhibits a pseudospherical morphology as shown in Fig. 3.20 however these particles begin to form agglomerates when the temperature within the thermal treatment increases up to 800 °C. After a thermal treatment at 800 °C the size of the particles has changed to micro-spheroid as can be clearly seen in figure. The grain

Fig. 3.19 SEM image of nanoparticles of ZrAl$_2$O$_3$ sample

Fig. 3.20 SEM image of nanoparticles of Bi_2O sample

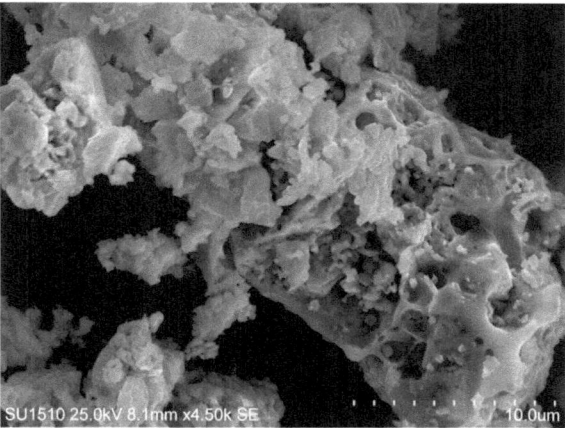

sizes estimated from SEM observations were different from those done by means of Scherrer's equation [20].

AgZnO

In order to investigate the morphology of the obtained sample, typical SEM image of Ag doped ZnO nanoparticles is shown in Fig. 3.21. The image reveals that most of them are irregularly shaped and relatively spherical with dimensions. The particles in this sample have relatively a sphere-like morphology and the nanoparticles were composed of agglomerates of Ag–ZnO particles [7, 21].

Ce–MgO

The SEM images for the Ce–MgO sample are shown in Fig. 3.22. From the SEM images the particle sizes of the nanocrystals were matched with PXRD, which is in quite accordance with the reported value. It is also clear that the synthesized MgO

Fig. 3.21 SEM image of nanoparticles of AgZnO sample

Fig. 3.22 SEM image of
nanoparticles of Ce–MgO
sample

sample is very porous in nature and when it is doped with Ce, the porosity increases
with pores and open voids [11].

Zn–MgO

In Fig. 3.23, the Zn doped MgO nanoparticles are agglomerated. This may be due
to the defects created by Zn doping. As the dopant concentration increases, the
agglomeration of particles takes place and hence particle size increases as compared
to the pure MgO nano particles [22].

Zn–MgO (Sol–Gel)

The morphological and structural studies were investigated using scanning electron microscopy and displayed in Fig. 3.24 for Zn doped MgO. These micrographs

Fig. 3.23 SEM image of
nanoparticles of Zn–MgO
sample

exhibited the formation of nanoparticles of Zn doped MgO. It was noted that the agglomeration increases with increasing Zn concentration [23].

NiO

The SEM images for the NiO sample are shown in Fig. 3.25. From the SEM images the particle sizes of the nanocrystals were matched with PXRD, which is in quite accordance with the reported value. It is also clear that the synthesized NiO sample is very porous and voids in nature. It depends on the synthesis method [12].

Fig. 3.24 SEM image of nanoparticles of Zn–MgO (sol–gel) sample

Fig. 3.25 SEM image of nanoparticles of NiO sample

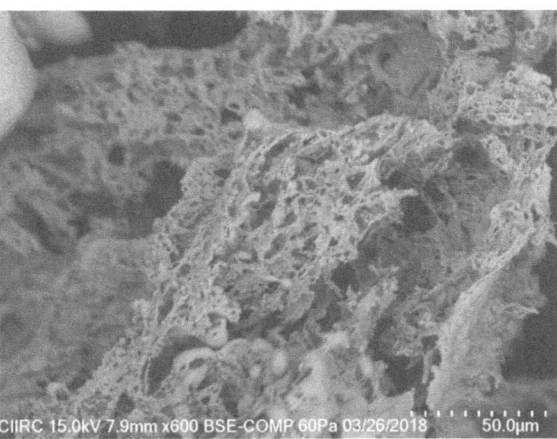

3.3 Conclusion

Metal oxide nanoparticles have been thoroughly analyzed in terms of their structure, morphology, and content, all of which demonstrate how significant a role they play in improving thermoluminescence applications. The effects of crystalline quality and particle size on TL efficiency are confirmed by XRD and SEM data, and the surface chemistry and functional group interactions are shown by FTIR spectra. The observed differences in phase purity and agglomeration across various materials highlight how crucial exact synthesis conditions are to achieving the most effective TL function. All things considered, these characterizations are crucial for creating sophisticated TL substances for radiation detection that are more accurate and dependable.

References

1. Azorin J (2014) Preparation methods of thermoluminescent materials for dosimetric applications: an overview. Appl Radiat Isot 83:187–191
2. Akhtar K, Khan SA, Khan SB, Asiri AM (2018) Scanning electron microscopy: principle and applications in nanomaterials characterization. Springer International Publishing, pp 113–145
3. Ossai CI, Raghavan N (2018) Nanostructure and nanomaterial characterization, growth mechanisms, and applications. Nanotechnol Rev 7(2):209–231
4. Kumari L, Li W, Wang D (2008) Monoclinic zirconium oxide nanostructures synthesized by a hydrothermal route. Nanotechnology 19(19):195602
5. Shtyka O, Maniukiewicz W, Ciesielski R, Kedziora A, Shatsila V, Sierański T, Maniecki T (2021) The formation of Cr–Al spinel under a reductive atmosphere. Materials 14(12):3218
6. Kim M (2008) Mixed-metal oxide nanopowders by liquid-feed flame spray pyrolysis (LF-FSP): synthesis and processing of core-shell nanoparticles. Doctoral dissertation
7. Shah AH, Basheer Ahamed M, Manikandan E, Chandramohan R, Iydroose M (2013) Magnetic, optical and structural studies on Ag doped ZnO nanoparticles. J Mater Sci Mater Electron 24:2302–2308
8. Umavathi S, AlSalhi MS, Devanesan S, Kadhiravan S, Gopinath K, Govindarajan M (2020) Synthesis and characterization of ZnO and Ca–ZnO nanoparticles for potential antibacterial activity and plant micronutrients. Surf Interfaces 21:100796
9. Varshney D, Dwivedi S (2015) On the synthesis, structural, optical and magnetic properties of nano-size Zn–MgO. Superlattices Microstruct 85:886–893
10. Sharma A, Arya S, Singh B, Prerna, Tomar A, Singh S, Sharma R (2020) Sol-gel synthesis of Zn doped MgO nanoparticles and their applications. Integr Ferroelectr 205(1):14–25
11. Hattab M, Hassen SB, Spriano S, Ferraris S, Cernea M, Amor YB (2024) Ce-doped MgO films on AZ31 alloy substrate for biomedical applications: preparation, characterization and testing. Biomed Mater 19(2):025013
12. Rahdar A, Aliahmad M, Azizi Y (2015) NiO nanoparticles: synthesis and characterization
13. Prakash T, Neri G, Bonavita A, Ranjith Kumar E, Gnanamoorthi K (2015) Structural, morphological and optical properties of Bi-doped ZnO nanoparticles synthesized by a microwave irradiation method. J Mater Sci Mater Electron 26:4913–4921
14. Fouladi-Fard R, Aali R, Mohammadi-Aghdam S, Mortazavi-Derazkola S (2022) The surface modification of spherical ZnO with Ag nanoparticles: a novel agent, biogenic synthesis, catalytic and antibacterial activities. Arab J Chem 15(3):103658
15. Salavati-Niasari M, Dadkhah M, Davar F (2009) Pure cubic ZrO_2 nanoparticles by thermolysis of a new precursor. Polyhedron 28(14):3005–3009

16. Farsi H, Moghiminia S, Roohi A, Hosseini SA (2014) Preparation, characterization and electrochemical behaviors of Bi_2O_3 nanoparticles dispersed in silica matrix. Electrochim Acta 148:93–103
17. Alaei M, Rashidi AM, Bakhtiari I (2014) Preparation of high surface area ZrO_2 nanoparticles. Iran J Chem Chem Eng (IJCCE) 33(2):47–53
18. Wu CH, Huang CN, Sun C, Kuan C, Shen P (2011) Directional diffusion-controlled development of spinel interlayer between zinc-orthosilicate glaze and alumina. Ceram Int 37(6):1801–1811
19. Pardo-Tarifa F (2017) Synthesis and characterization of novel $Zr–Al_2O_3$ nanoparticles prepared by microemulsion method and its use as cobalt catalyst support for the CO hydrogenation reaction. Synth Catal Open Access 2(2):1–9
20. Jalalah M, Faisal M, Bouzid H, Park JG, Al-Sayari SA, Ismail AA (2015) Comparative study on photocatalytic performances of crystalline α- and β-Bi_2O_3 nanoparticles under visible light. J Ind Eng Chem 30:183–189
21. Lu YF, Lan WH, Wang MC, Shih MC, Kuo HH, Feng DJY, Chiu YJ, Hung YJ, Yang CF (2018) Carrier concentration of calcium zinc oxide with different calcium contents deposited through spray pyrolysis. Microsyst Technol 24:4267–4272
22. Parvizi E, Tayebee R, Koushki E (2019) Mg-doped ZnO and Zn-doped MgO semiconductor nanoparticles; synthesis and catalytic, optical and electro-optical characterization. Semiconductors 53:1769–1783
23. Mansoor S, Shahid S, Ashiq K, Alwadai N, Javed M, Iqbal S, Fatima U, Zaman S, Sarwar MN, Alshammari FH, Elkaeed EB, Awwad NS, Ibrahium HA (2022) Controlled growth of nanocomposite thin layer based on Zn-doped MgO nanoparticles through sol-gel technique for biosensor applications. Inorg Chem Commun 142:109702

Chapter 4
Thermoluminescence Properties of Metal Oxide Nanomaterials

Abstract The thermoluminescence characteristics of metal oxide nanoparticles are examined in this chapter with an emphasis on the way they detect ionizing radiation. Charge carriers that are trapped in a material's crystal lattice and that release illumination are known as thermoluminescence. The chapter describes thermoluminescence experiment arrangements, presents results from several metal oxide nanoparticles like BiZnO, CaZnO, CeMgO, ZnMgO, and ZrO_2 and talks about the glow curve properties of these nanoparticles. It highlights how ZrO_2 performs and is more sensitive in detecting radiation application.

4.1 Introduction

Thermoluminescence is a technique that uses a material's emitted light upon heat to detect and quantify radiation that is ionizing. The idea is to use ionizing radiation to encase charge carriers, including electrons or holes, inside the crystal structure of a luminous substance, like calcium sulphate or lithium fluoride. These trapped charge carriers build up within the material at particular energy levels. The trapped carriers are released and reunite with luminous centres when the material is heated later on, producing light in the process. It is possible to estimate the radiation dosage accurately because the intensity of the light emitted is proportionate to the quantity of radiation the material has absorbed [1]. Thermoluminescence is useful in identifying radiation owing to its great sensitivity and ability to assess a variety of radiation kinds and dosages. It is frequently employed in dosimetry for irradiation surveillance in the workplace, environment, and medicine. The technique is useful for guaranteeing compliance and security in a variety of applications as it offers accurate dose estimates and can identify minimal radiation levels. The fact that thermoluminescence materials may also be used to date geological and archaeological samples adds to their versatility concerning practical and scientific uses [2].

The thermoluminescence measurements experimental setup consists of many essential parts and steps. The process of preparing a sample involves subjecting a thermoluminescent substance, like calcium sulfate or lithium fluoride, to a predetermined amount of radiation that is ionizing. It is essential to heat the sample carefully, this is usually accomplished using a furnace that warms the material precisely or with a thermoluminescence reader. Light detecting is the process of detecting light output that is proportionate to the quantity of radiation absorbed. This is frequently done with a photomultiplier tube or a photodiode. To calculate the radiation dosage, statistical analysis matches the light brightness to a calibration curve. Accurate readings need backdrop adjustment and proper calibration [3].

Evaluating the photon emissions sensitivity of produced metal oxide nanoparticles to temperature stimulation is a necessary step in evaluating their thermoluminescent capabilities. To cause charge carrier entrapment, the nanomaterials, such as ZnO or TiO_2, are first exposed to known levels of ionizing radiation. The substances are then heated under controlled circumstances with the aid of a thermoluminescence reader. The intensity and efficiency of luminescence are measured by analyzing the emitted light, which is picked up by a photomultiplier tube or other such device. The material's efficacy in detecting radiation and dosimetry is shown by key qualities such as sensitivity, maximum temperature of luminescence, and glow curve features [4].

4.2   Thermoluminescence Studies

Thermoluminescence (TL) also known Thermally Stimulated Luminescence (TSL) is type of delayed phosphorescence, where the photon energy is released when the crystalline substance is heated from low temperature to high temperature. In thermoluminescence heat is not the primary source of energy, only the trigger for the release of energy that originally came from another source. TL occurs as a result of heating, when electrons in metastable state make radiative transitions. The release of the trapped charge carriers by thermal stimulation and their recombination produces the light called Thermoluminescence [5].

Upon irradiating the solid with ionizing radiations (X-rays, gamma rays, beta rays, alpha rays or energetic ion beams) some electrons gain sufficient energy to be raised from the valence to the conduction band from which they may subsequently be trapped at defect centers [6]. When the solid is heated these trapped electrons can gain enough thermal energy to escape from the traps back to the conduction band. From here they may make direct transitions back to the valence band or, alternatively, they may become retrapped or may combine with trapped holes. If the electron trap energy levels are close to the conduction band, then this thermal untrapping may occur at ambient temperatures and if recombination with holes leads to light emission, then the resulting process is phosphorescence [7].

Figure 4.1 shows the thermoluminescence glow curves of combustion synthesized BiZnO, CaZnO, CeMgO, ZnMgO and ZrO_2 gamma rayed for doses in the range 0.087–5.822 kGy. The glow curves of BiZnO clearly shows prominent and well separated glows with peaks around 160 °C (T_{g1}) and 374 °C (T_{g2}). The glow curves of CaZnO show prominent glow peat at 188 °C (T_{g1}). The glow curves of CeMgO clearly shows prominent and well separated glows with peak around 171 °C (T_{g1}) and a shoulder peak at 232 °C (T_{g2}). The glow curves of ZnMgO show prominent glow peat at 181 °C (T_{g1}). However, the glow curves of ZrO_2 show non prominent glow peat at 250 °C (T_{g1}). TL intensity increases almost linearly with increase in gamma ray dose for combustion synthesized material upto 60 min irradiation time and then decreases may due to quenching [8].

The peak position of TL glows (T_{g1} and T_{g2}) is almost steady for the entire dose range and might be due to the occupancy of deep traps and it may be attributed to disorganization of the initial energy levels as a result of increasing dose.

The trapping parameters known as kinetic parameters were calculated according to glow curve shape method using computerized glow curve deconvolution technique and results are tabulated in Table 4.1. TL glow peak depends on various parameters such as history of the samples, heat-treatment given to the samples prior to irradiation, physical nature of the sample, impurity content of the sample, nature and amount of dose given to the sample, temperature at which irradiation as well as TL measurements are made, environment of the sample while irradiation as well as TL measurements, rate of heating the sample etc. On irradiation different kinds of radicals and colour centers such as SO_4^-, SO_3^-, SO_2^-, O^-, F^-, V_k^+, α_2^{3+} (electron captured at an anion vacancy), etc., can form and act as a source of trapped electrons or holes [8].

4.3 Conclusion

The thermoluminescence (TL) characteristics of metal oxide nanoparticles are thoroughly examined in this chapter, emphasizing the materials' use in radiation detection. According to the research, ZrO_2 exhibits better thermoluminescence properties than the other substances that were examined, including BiZnO, CaZnO, CeMgO, and ZnMgO. ZrO_2 has a robust linear response to rising gamma-ray doses and a noticeable glow peak at 250 °C, showing its outstanding sensitivity and consistent effectiveness for irradiation dosimetry. The resilience of the material is demonstrated by its ability to retain constant peak locations and effective luminescence even at larger dosages. ZrO_2 is a promising material for applications requiring accurate and long-lasting radiation monitoring, as indicated by the estimated kinetic characteristics. All things considered, ZrO_2 sticks out as an outstanding material for thermoluminescent uses providing notable advantages in accuracy and operational efficiency.

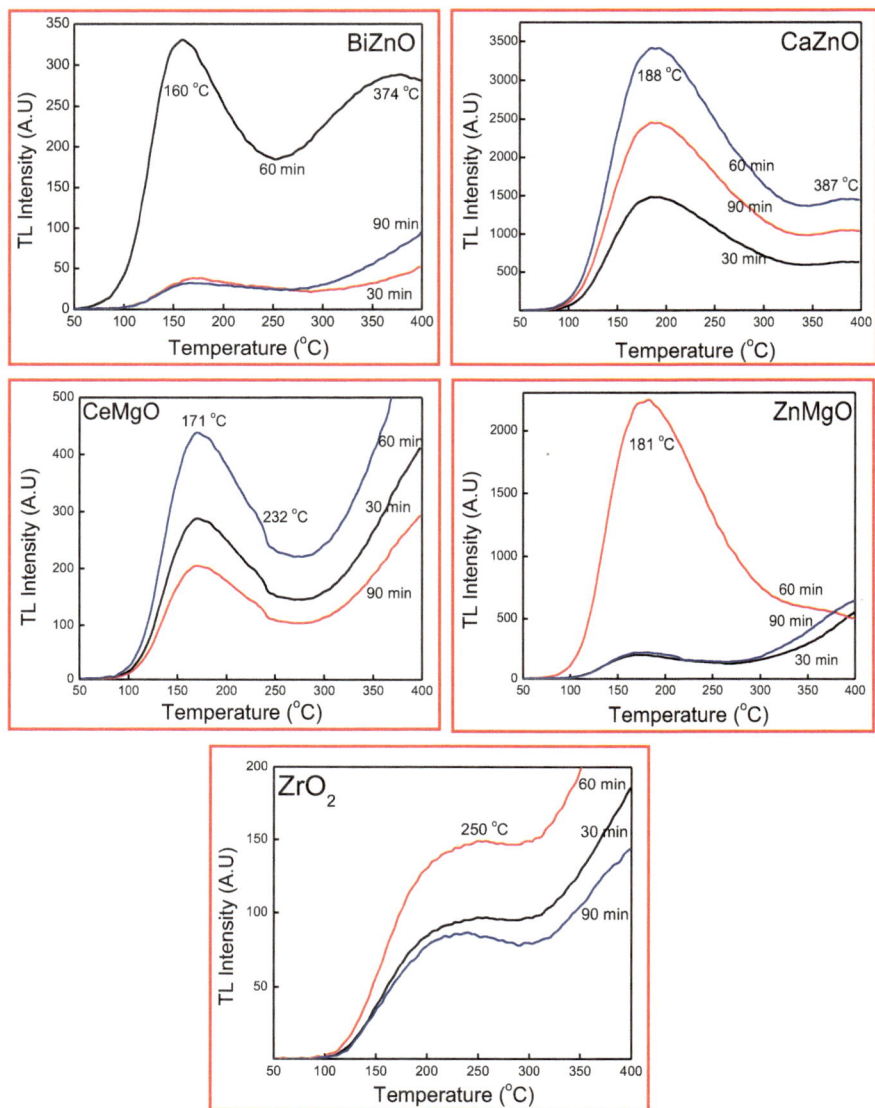

Fig. 4.1 Thermoluminescence curves of combustion synthesized BiZnO, CaZnO, CeMgO, ZnMgO and ZrO$_2$

Table 4.1 Kinetic parameters obtained by using the glow curve shape method (modified by Chen) for γ-rayed samples

Sample	T_m (°C)	μ_g	b	E_τ (eV)	E_δ (eV)	E_ω (eV)	E_a (eV)	n_o (cm^{-3})	s (s^{-1})
BiZnO	160	0.491	2	0.683	0.723	0.706	0.704	6.42×10^3	9.28×10^8
CaZnO	188	0.490	2	0.550	0.620	0.590	0.592	2.69×10^4	1.6×10^6
CeMgO	171	0.505	2	1.066	1.105	1.091	1.087	5.46×10^3	1.84×10^{10}
ZnMgO	181	0.503	2	1.063	1.11	1.093	1.082	5.41×10^3	1.82×10^{10}
ZrO$_2$	250	0.390	1	0.101	0.182	0.124	0.136	4.29×10^4	5.00×10^1

References

1. Bos AJ (2017) Thermoluminescence as a research tool to investigate luminescence mechanisms. Materials 10(12):1357
2. Mejdahl V, Wintle AG (2020) Thermoluminescence applied to age determination in archaeology and geology. In: Thermoluminescence and thermoluminescent dosimetry. CRC Press, pp 133–190
3. Izewska J, Rajan G (2005) Radiation dosimeters. In: Radiation oncology physics: a handbook for teachers and students, vol 36, pp 71–99
4. Nieto JA (2016) Present status and future trends in the development of thermoluminescent materials. Appl Radiat Isot 117:135–142
5. Murthy KVR (2014) Thermoluminescence and its applications: a review. In: Defect and diffusion forum, vol 347. Trans Tech Publications Ltd., pp 35–73
6. Fink D (ed) (2004) Fundamentals of ion-irradiated polymers, vol 63. Springer Science & Business Media
7. Morehead FF (1958) Electron traps and the electroluminescence brightness and brightness waveform. J Electrochem Soc 105(8):461
8. McKeever SW (1980) On the analysis of complex thermoluminescence. Glow-curves: resolution into individual peaks. Phys Stat Sol (a) 62(1):331–340

Chapter 5
Integration of ZrO$_2$ Nanomaterials into Scintillation Detectors

Abstract This chapter explores the integration of ZrO$_2$ nanomaterials into plastic scintillation detectors, focusing on enhancing radiation detection capabilities. Plastic scintillators, composed of organic polymers, are widely used in radiation detection due to their efficiency, lightweight nature, and versatility. The addition of ZrO$_2$ nanoparticles to these scintillators significantly improves their performance by increasing the refractive index, leading to enhanced light collection and reduced scattering losses. However, challenges such as achieving uniform nanoparticle distribution within the scintillator matrix are addressed by employing advanced dispersion techniques and surface modifications. The chapter details the synthesis of surface-modified ZrO$_2$ nanoparticles and their incorporation into plastic scintillators. Various characterization techniques, including X-ray diffraction (XRD), scanning electron microscopy (SEM), Fourier transform infrared spectroscopy (FTIR), and dynamic light scattering (DLS), are employed to analyze the structural, morphological, and optical properties of the modified ZrO$_2$ nanoparticles. The results demonstrate that surface modifications lead to smaller crystallite sizes and improved dispersion stability, making the modified nanoparticles more suitable for enhancing the performance of plastic scintillators. The study concludes by highlighting the improved scintillation efficiency and durability of ZrO$_2$-integrated plastic scintillators, making them promising candidates for advanced radiation detection applications.

5.1 Introduction

Plastic scintillators are essential components in radiation detection devices due to their efficient transformation of electromagnetic radiation into visible light. These detectors are composed of organic polymers, such as polystyrene or polyvinyl toluene, that are coated with fluorescent compounds that produce light when activated by incoming radiation. The emitted light is then collected by light detectors, which include photomultiplier tubes (PMTs) or silicon photomultipliers (SiPMs) and

© The Author(s), under exclusive license to Springer Nature Switzerland AG 2024
R. Naik et al., *Advances in Space Radiation Detection*,
SpringerBriefs in Molecular Science, https://doi.org/10.1007/978-3-031-74551-5_5

transformed into an electrical signal for examination. Plastic scintillators are popular due to their lightweight, resilience, and adaptability, which allows them to be created in a variety of forms and sizes to suit various uses [1, 2].

They have excellent detecting efficiency and rapid reaction times, which makes them ideal for application in particle physics research, imaging for medicine, and environmental radiation monitoring. Adding ZrO$_2$ nanoparticles to plastic scintillators can significantly improve their performance. ZrO$_2$ nanoparticles increase the scintillator's refractive index, improving light collecting and decreasing scattering losses. This integration has resulted in significant increases in light production and energy resolution, allowing for more precise and dependable radiation detection. The improved scintillation efficiency and endurance make the modified scintillators useful for advanced studies as well as practical detection of radiation purposes [3, 4].

5.2   Challenges in Uniformly Distributing Nanomaterials Within Plastic Scintillators

The uniform distribution of nanoparticles throughout plastic scintillators involves many obstacles. Dispersion nanoparticles agglomerate owing to strong van der Waals forces, resulting in unequal distribution across the scintillator matrix. This can lead to irregular scintillation qualities and decreased performance. Mixing techniques provide additional problems; establishing a uniform blend of nanoparticles with a matrix of polymers necessitates careful control over mixing operations to minimize clumps and guarantee equal distribution [5]. Different chemical qualities can impact the dispersion's durability and the connection between the nanomaterials and the scintillator matrix, therefore compliance between the nanoparticles and the polymer is also an issue. Furthermore, the distribution of nanoparticles can be affected by processing parameters including temperature and shear pressures during polymerization or extrusion, which may result in inhomogeneities. To overcome these obstacles and maintain outstanding efficiency in plastic scintillators, evenly dispersed and precise choice of materials are necessary, as are optimum manufacturing conditions. Advanced techniques in nanomaterial dispersion are also needed [6].

5.3   Development of a Process for 3-D Uniform Distribution of Nanomaterials

There are numerous crucial elements when creating a technique to achieve a 3-D uniform dispersion of nanomaterials. Dispersal to ensure equal dispersion throughout the polymer matrix and prevent nanoparticle accumulation, methods like high-shear mixing or ultrasonic mixing are employed. Surface The process of functionalizing nanoparticles improves their compatibility with polymers, leading to

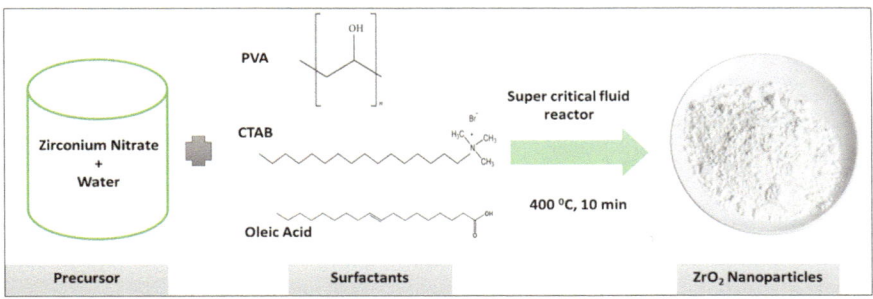

Fig. 5.1 Methodology for synthesizing surface modified ZrO_2 nanoparticles

increased stability and homogeneity. Enhanced Processing Consistent nanoparticle dispersion may be maintained during polymerization by controlling variables like temperature and mixing speed. Furthermore, sophisticated production techniques like 3-D printing or electrospinning may be used to evenly include nanoparticles throughout the scintillator. When combined, these tactics guarantee an even distribution, improving the end product's reliability as well as efficiency [7].

5.4 Synthesis of Surface Modified ZrO_2

To synthesize surface modified ZrO, 0.1 mol of zirconium nitrate and 1 wt % surfactants was added into 20 ml of D.I. water and stirred for 30 min, after complete dissolution the solution was transferred into super critical fluid (SCF) reactor and kept at 400 °C for 10 min. After reaction complete the reactor was quenched in cold water [8]. The obtained product was taken out and washed in water and methanol repeatedly then dried at 60 °C overnight to get powder as shown in Fig. 5.1.

5.5 Preparation of Plastic Scintillators

The plastic scintillators were prepared by using the methodology as shown in Fig. 5.2. Initially 2.5 g of polystyrene beads and 12.5 mg PBD were dissolved in 25 ml of THF solution then, 1 wt % of pure ZrO_2 was slowly added during sonication. The sonication was carried out for 30 min to completely disperse the nanoparticles as shown in figure. The same procedure was repeated for Oleic acid-ZrO_2, CTAB-ZrO_2 and PVA-ZrO_2 to prepare corresponding plastic scintillators [9].

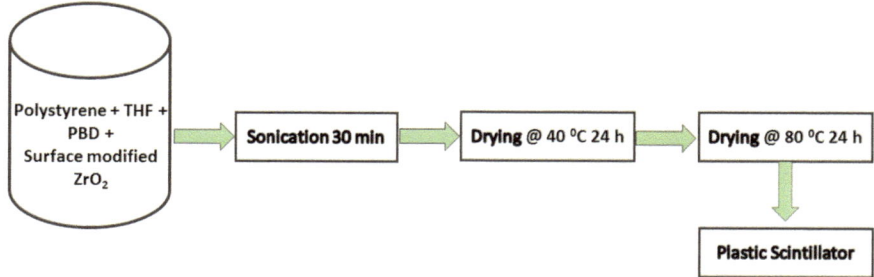

Fig. 5.2 Methodology of plastic scintillator preparation

5.6   Method of Analysis/Methodology

The prepared samples are characterized by various techniques, crystal structures and phase purity of the as synthesized samples were carried out using powder X-Ray diffractometer (PXRD) (Rigaku Ultima-IV) ranging from 10 to 80° in 2-theta with 0.02 steps/s using Cu-Kα as radiation source ($\lambda = 0.15406$ nm) at room temperature. The morphology and elemental analyses of the prepared samples were analysed by scanning electron microscopy (SEM) SU1510 and energy dispersive X-Ray spectroscopy (EDX), respectively. The UV–Vis spectra of the samples were carried out in Perkin–Elmer spectrometer. Fourier transform infrared spectroscopy of the samples was carried out in Perkin–Elmer UATR mode spectrometer.

5.7   Results and Discussion

5.7.1   Crystal Structure Analysis

Figure 5.3 shows the PXRD patterns of the ZrO$_2$ nanoparticle and surface modified ZrO$_2$ nanoparticle produced using the supercritical hydrothermal method as reported by Muttin & Lagashetty [8]. All the products mainly showed monoclinic structure (JCPDS No. 83-0944). The lattice parameters did not show any remarkable difference among the products. Therefore, the crystallinity of the monoclinic ZrO$_2$ appears to be independent of the surface modification, even though differences in the crystallite size between products prepared with and without addition of surface modifiers were apparent.

The crystallite size of the products decreased with addition of the surface modifiers when compared unmodified ZrO$_2$. The growth of (111) plane reduced with the addition of modifiers [10, 11]. The average crystallite size calculated by Debye Scherrer's formulae is shown in Table 5.1. The decrease in crystallite size due to the addition of surface modifiers is likely caused by the modifiers acting as growth inhibitors, disrupting the normal growth process of the ZrO$_2$ nanoparticles. These

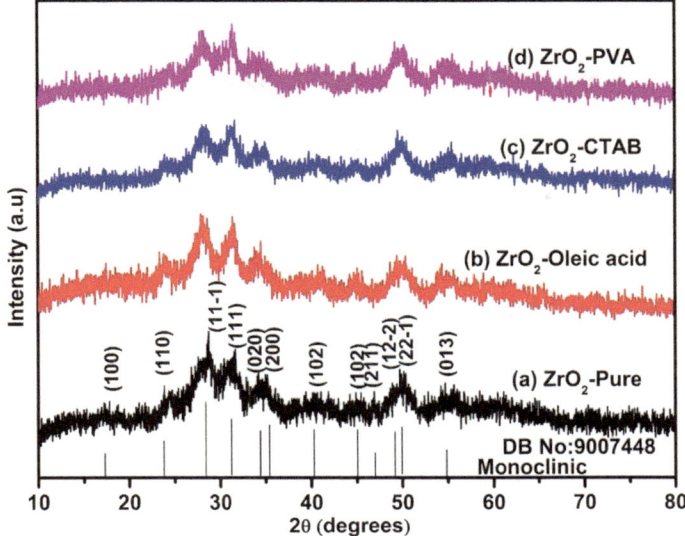

Fig. 5.3 XRD patterns of **a** ZrO$_2$, **b** oleic acid-ZrO$_2$, **c** CTAB-ZrO$_2$ and **d** PVA-ZrO$_2$ [8]

Table 5.1 Average crystallite size calculated by Debye Scherrer's formula

Name of the material	Crystallite size (nm)
ZrO$_2$-pure	39
ZrO$_2$-OA	30
ZrO$_2$-CTAB	32
ZrO$_2$-PVA	35

inhibitors attach to the surface of the growing crystals, particularly along the (111) plane, slowing down their growth. This effect can enhance surface-related properties, improving performance in applications requiring high surface areas, such as catalysis, sensors, or energy storage.

5.7.2 SEM Analysis

Muttin & Lagashetty [8] reported SEM images of ZrO$_2$, Oleic acid-ZrO$_2$, CTAB-ZrO$_2$ and PVA-ZrO$_2$. In all the above images the size of the particles were found to be in the range of nano size [12]. Scanning Electron Microscopy (SEM) images that depict the morphological characteristics of ZrO$_2$ nanoparticles and their surface modifications using different surfactants or capping agents. These images provide valuable insights into the size, shape, and distribution of the nanoparticles, which are crucial factors influencing their physical and chemical properties.

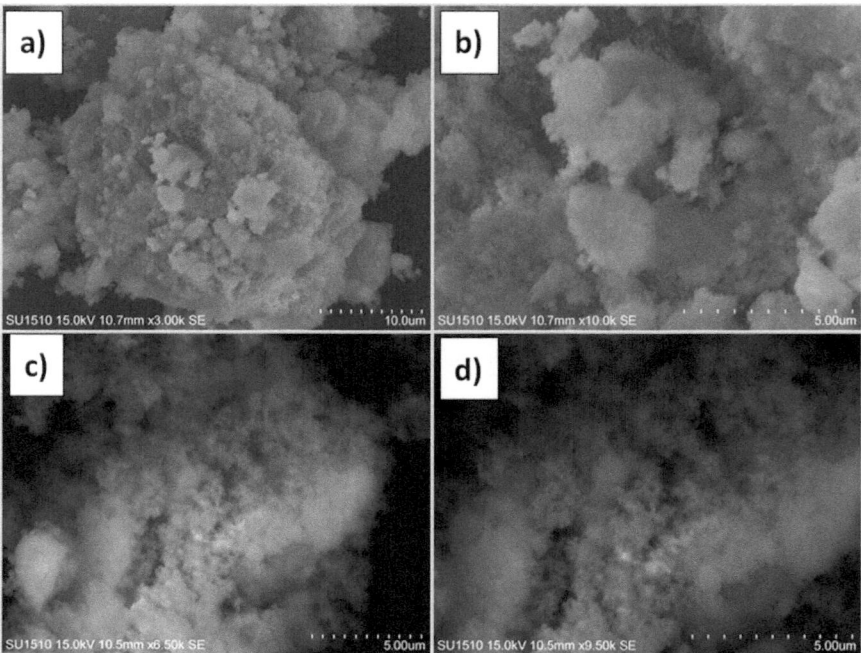

Fig. 5.4 SEM images of **a** and **b** ZrO$_2$ and **c** and **d** oleic acid-ZrO$_2$

Figure 5.4: ZrO$_2$ Nanoparticles: This image shows the morphology of pure ZrO$_2$ nanoparticles. The particles are uniformly distributed and exhibit a spherical shape. The size of the nanoparticles is in the nanoscale range, typically between 20 and 50 nm. The absence of agglomeration suggests good dispersion of the nanoparticles, which is essential for maintaining their high surface area and reactivity [8].

Oleic Acid-ZrO$_2$ Nanoparticles: In this image, ZrO$_2$ nanoparticles are surface-modified with oleic acid. The presence of oleic acid is evident from the slight increase in particle size compared to pure ZrO$_2$. The oleic acid acts as a capping agent, preventing the nanoparticles from aggregating and thus maintaining their nanoscale size. This modification also introduces hydrophobicity to the surface, which could be beneficial in specific applications where water resistance is desired [8].

Figure 5.5: CTAB-ZrO$_2$ Nanoparticles: The SEM image of ZrO$_2$ nanoparticles modified with CTAB (cetyltrimethylammonium bromide) shows a more pronounced effect on the particle morphology. CTAB, being a cationic surfactant, not only prevents agglomeration but also aids in the formation of well-defined nanostructures. The particles appear to be more uniform and slightly larger than the unmodified ZrO$_2$ due to the surfactant layer. This modification can enhance the nanoparticle's interaction with specific substrates, making them suitable for catalysis or adsorption processes [8].

Figure 5.6: PVA-ZrO$_2$ Nanoparticles: The SEM image of ZrO$_2$ nanoparticles capped with PVA (polyvinyl alcohol) shows a distinct morphology compared to the other modifications. PVA creates a more extensive polymer network around the

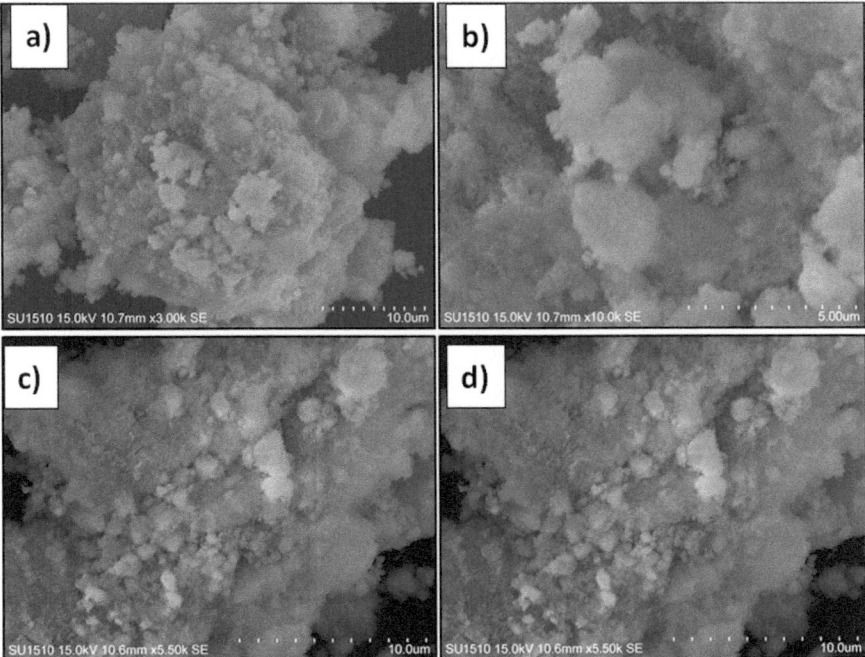

Fig. 5.5 SEM images of **a** and **b** ZrO$_2$ and **c** and **d** CTAB-ZrO$_2$

nanoparticles, leading to a slight increase in particle size. The nanoparticles remain in the nanoscale range, and the PVA coating provides excellent stability in aqueous environments, making these nanoparticles ideal for biomedical applications or in environments where moisture is present [8].

In all the images, the particle sizes are consistently within the nanometer range, confirming the successful synthesis and modification of ZrO$_2$ nanoparticles. The choice of capping agent significantly influences the morphology and stability of the nanoparticles, which in turn affects their performance in various applications.

5.7.3 FTIR Analysis

The interactions between the surface modifiers and the nano-particles were assessed by FT-IR measurements as shown in Fig. 5.7 [8]. All the ZrO$_2$ products showed two distinctive bands with two peaks around 750 and 500 cm, which are assigned to the Zr–O modes in monoclinic ZrO$_2$. The Oleic acid-ZrO showed a weak band at 2962 cm^{-1}, which is assigned to the asymmetric stretching mode of –CH$_3$ in oleic acid (OA). The spectrum some bands at 2919, 2849 and 1453 cm^{-1}, which are assigned to the asymmetric stretching and symmetric stretching and scissoring bending –CH$_2$ group in the oleic acid. The absorption peaks at 1537 and 1408 cm^{-1}

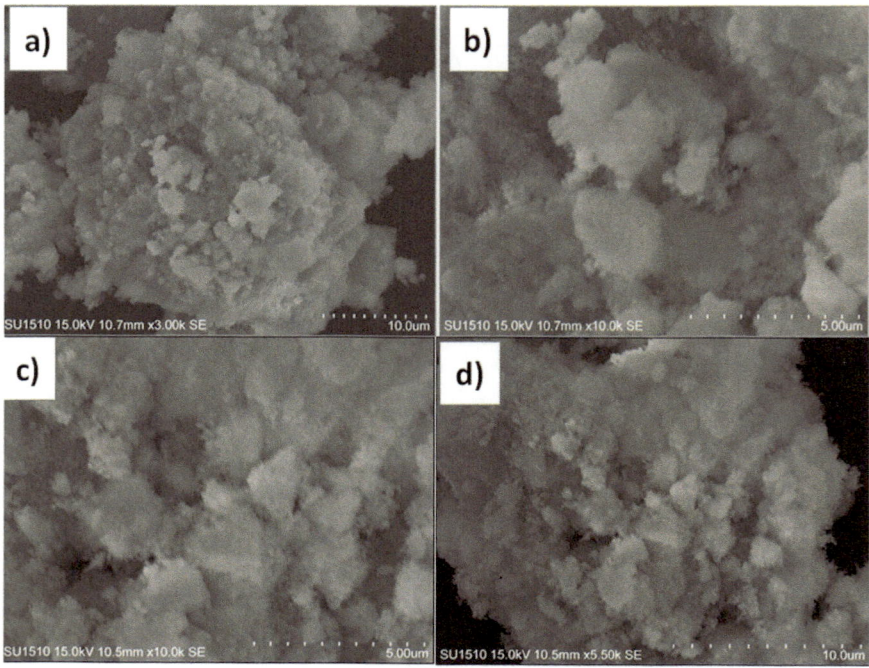

Fig. 5.6   SEM images of **a** and **b** ZrO$_2$ and **c** and **d** PVA-ZrO$_2$

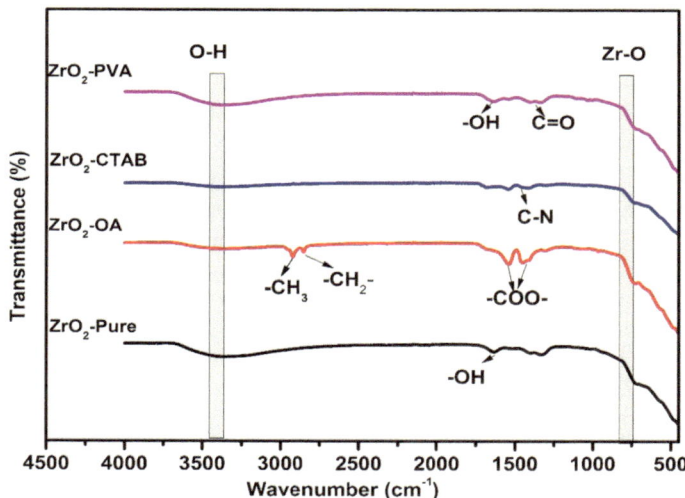

Fig. 5.7   FTIR analysis of ZrO$_2$, oleic acid-ZrO$_2$, CTAB-ZrO$_2$ and PVA-ZrO$_2$ [8]

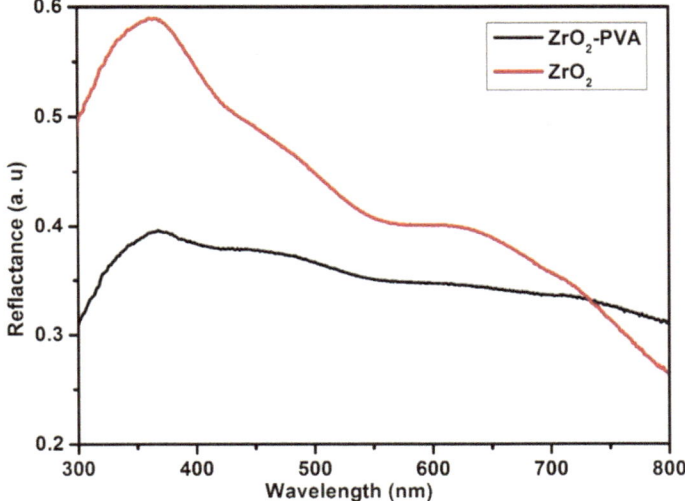

Fig. 5.8 Diffuse reflectance spectra of **a** ZrO$_2$ and **b** PVA-ZrO$_2$ [8]

confirms symmetric and asymmetric stretching vibrations of COO– groups in the oleic acid. At 1460 cm^{-1} the low absorption peak indicates the stretching vibrations of C–N bonds present in the cationic CTAB molecule [10].

5.7.4 UV–Vis Analysis

Figure 5.8 shows diffused reflectance spectroscopy of ZrO$_2$ and PVA-ZrO$_2$ nanoparticles. The maximum reflectance was observed at 370 nm [8]. The reflectance of pure ZrO$_2$ was found to be higher compared to PVA-ZrO$_2$. This may be due to the pure form of ZrO$_2$ without any surface modifications [13].

5.7.5 TGA Analysis

The TGA curves of ZrO$_2$ and PVA-ZrO$_2$ shown in Fig. 5.9. The amount of surface modifiers attached on to the surface of ZrO$_2$. The total weight loss was found to be 18%. There were three major weight losses were observed. The weight loss at temperature from 50 to 200 °C mainly due to adsorbed water molecules. Usually, PVA degrade below 300 °C itself but from the curve it is clear that after 300 °C also there is a significant amount of weight loss. This indicates formation coordination bond between Zr atom and molecules of PVA which is also confirmed by FTIR spectroscopy [14].

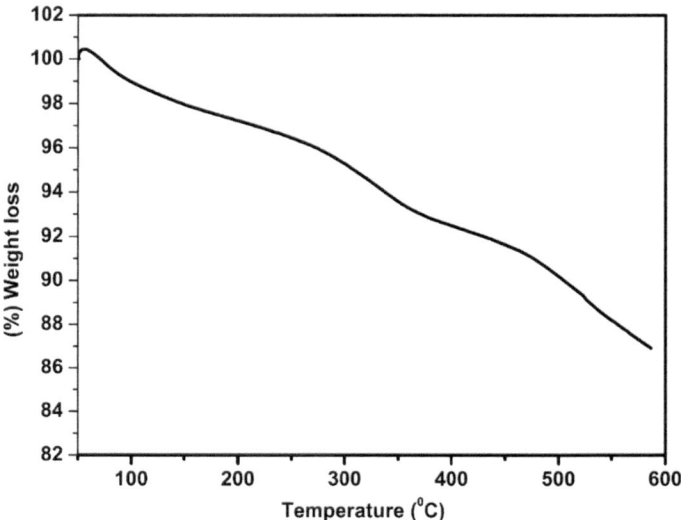

Fig. 5.9 TGA curve of **a** ZrO$_2$ and **b** PVA-ZrO$_2$

5.7.6 DLS Analysis

Dynamic light scattering was performed by Muttin & Lagashetty [8] carried out to determine surface charges and dispersion stability of the synthesized ZrO$_2$ and PVA-ZrO$_2$. Zeta potential was determined by knowing the mobility of colloidal particles suspended in a water by applying electric field. The hydrodynamic size and poly dispersion intensity was determined to confirm the surface functionalized moieties and dispersion stability of the synthesized ZrO$_2$ and PVA-ZrO$_2$. The zeta potential of ZrO$_2$ and PVA-ZrO$_2$ is shown in Fig. 5.10. The ZrO$_2$ surface modified with PVA shows 3 mV zeta potential whereas without modified ZrO$_2$ has + 3 mV. The PVA-ZrO$_2$ shows negative charge because of negative charges on the surface, which, confirms the modification of ZrO$_2$ with PVA [8, 15].

The hydrodynamic size and polydispersion intensity of both ZrO$_2$ and PVA-ZrO$_2$ is shown in Fig. 5.11. The average hydrodynamic size of pure ZrO$_2$ was found to be 178.7 nm with polydispersion intensity of 0.475. There were two peaks one at 200 nm

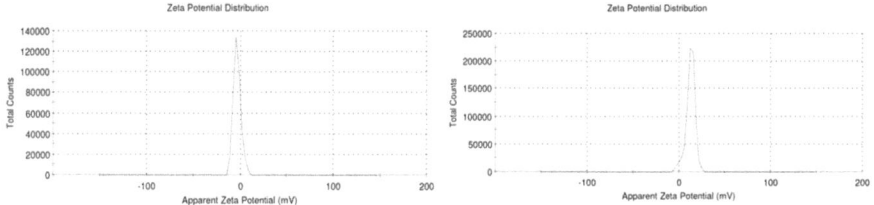

Fig. 5.10 Zeta potential of ZrO$_2$ (left) and PVA-ZrO$_2$ (right) [8]

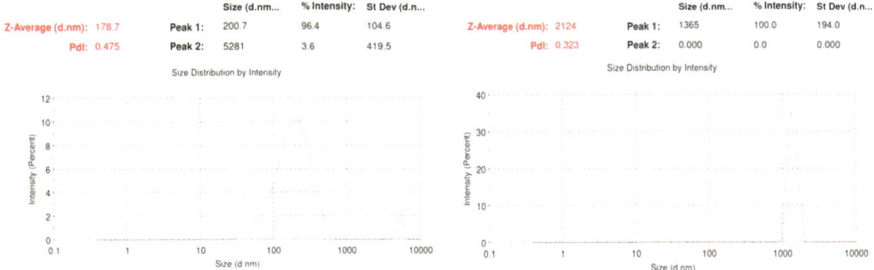

Fig. 5.11 Size distribution of ZrO₂ (left) and PVA-ZrO₂ (right) [8]

and another at 5281 nm which indicates the aggregation of pure ZrO_2 nanoparticles in other words polydispersion.

In PVA-ZrO_2 the hydrodynamic size was found to be 1365 nm with 0.323 polydispersion intensity. The higher hydrodynamic size probably due to denser surface attached PVA molecules and there was no other peak. Which, indicates there was no aggregation and hence no polydispersion occurred in PVA modified ZrO_2. More uniform size and much stable particles are suitable for plastic scintillator preparation. Therefore, PVA modified ZrO_2 were found to be more suitable in plastic scintillator preparation [16].

5.8 Plastic Scintillator Preparation

The plastic scintillator preparation flow chart is shown in Fig. 5.11. Initially, 2.5 g of polystyrene crystals and 12 mg of PBD were added into 25 ml of Tetrahydrofuran (THF) which, were taken in 50 ml glass beaker. Raised the temperature of hot plate to 40 °C and stirred for 15 min to complete the dissolution of polystyrene and PBD. Then, added slowly 25 mg of (1 wt %) ZrO_2 or PVA-ZrO_2 under sonication and continued for 30 min to disperse added NP's completely in the THF solution as shown in Fig. 5.12. The dispersed solution was kept at room temperature for 3 h then, transferred to hot air oven for drying at 40 °C for one day. After one day drying at 40 °C the solution becomes semi solid [17].

5.9 ZrO₂-PVA Plastic Scintillator

After complete drying (48 h at 75 °C), the semi solid becomes dry solid then the beaker was break down to get the dried solid plastic scintillator as shown in Fig. 5.13. For preparation of CTAB-ZrO_2 dispersed plastic scintillators CTAB-ZrO_2 was added instead of PVA-ZrO_2 [8]. Different shape and sized plastic scintillators were prepared as shown in Figs. 5.14, 5.15 and 5.16.

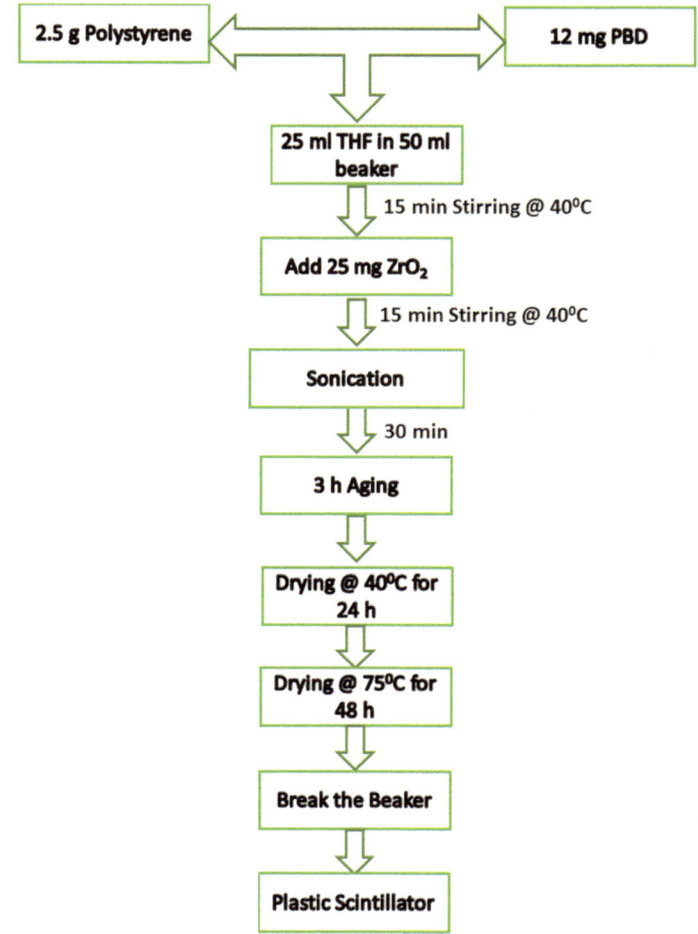

Fig. 5.12 Flow chart of plastic scintillator preparation

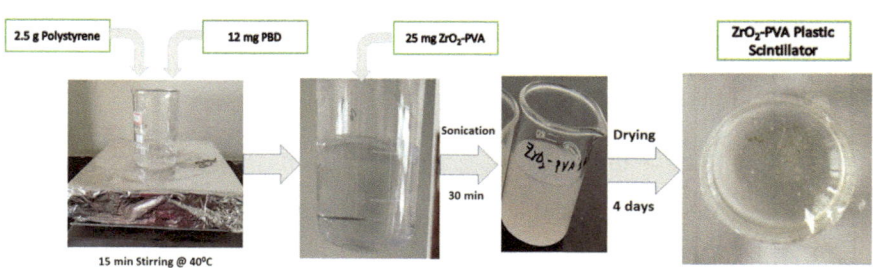

Fig. 5.13 Steps in plastic scintillator preparation

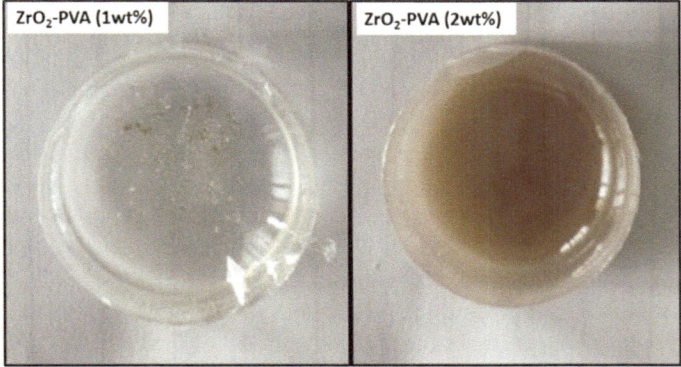

Fig. 5.14 Complete dried plastic scintillators

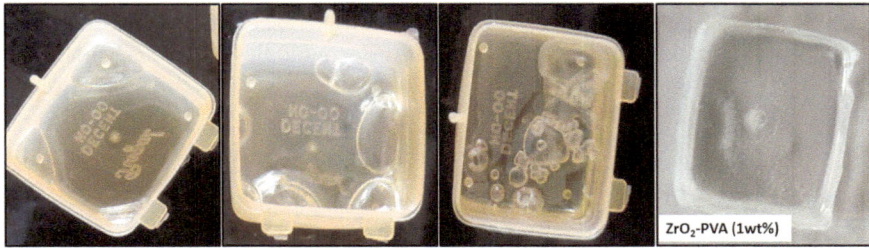

Fig. 5.15 Square shaped plastic scintillators

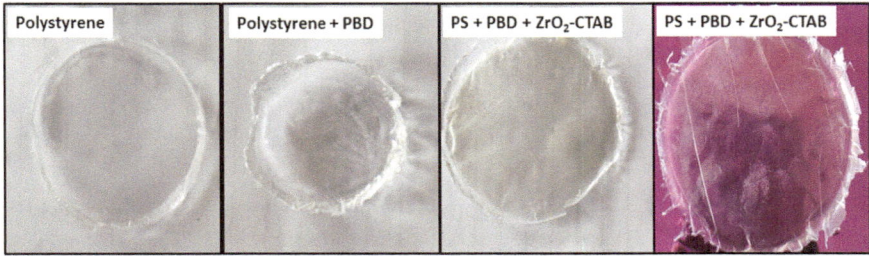

Fig. 5.16 Sheets of plastic scintillators

5.10   Conclusion

This chapter has demonstrated the significant advancements achieved by integrating ZrO_2 nanomaterials into plastic scintillation detectors. The incorporation of surface-modified ZrO_2 nanoparticles has proven to be an effective strategy for enhancing the performance of plastic scintillators. By increasing the refractive index of the scintillator material, these nanoparticles contribute to improved light collection efficiency and reduced scattering losses. The successful synthesis and characterization of these nanoparticles, utilizing techniques such as XRD, SEM, FTIR, and DLS, have provided valuable insights into their structural, morphological, and optical properties. The findings indicate that surface modifications result in smaller crystallite sizes and better dispersion stability, which are crucial for optimizing scintillator performance. Overall, the integration of ZrO_2 into plastic scintillators has led to enhanced scintillation efficiency and durability. These improvements make ZrO_2-integrated scintillators highly promising for advanced radiation detection applications, paving the way for more accurate and reliable detection technologies in various fields. Future work may focus on further optimizing the nanoparticle characteristics and exploring their performance in different radiation detection scenarios.

References

1. Hajagos TJ, Liu C, Cherepy NJ, Pei Q (2018) High-Z sensitized plastic scintillators: a review. Adv Mater 30(27):1706956
2. Stagliano M (2019) Silicon photomultiplier current and prospective applications in biological and radiological photonics
3. Renuka L, Anantharaju KS, Sharma SC, Nagabhushana H, Vidya YS, Nagaswarupa HP, Prashantha SC (2017) A comparative study on the structural, optical, electrochemical and photocatalytic properties of ZrO_2 nanooxide synthesized by different routes. J Alloy Compd 695:382–395
4. Toda A, Kishimoto S (2020) X-ray detection capabilities of plastic scintillators incorporated with ZrO_2 nanoparticles. IEEE Trans Nucl Sci 67(6):983–987
5. Braddock IHB (2023) Lead halide perovskite nanocomposite scintillators. Doctoral dissertation, University of Surrey
6. Wang B, Li P, Zhou Y, Deng Z, Ouyang X, Xu Q (2022) $Cs_3Cu_2I_5$ perovskite nanoparticles in polymer matrix as large-area scintillation screen for high-definition X-ray imaging. ACS Appl Nano Mater 5(7):9792–9798
7. Rozenberg BA, Tenne R (2008) Polymer-assisted fabrication of nanoparticles and nanocomposites. Prog Polym Sci 33(1):40–112
8. Muttin V, Lagashetty A (2022) Synthesis, characterization and studies of ZrO_2-PVA composites using super critical fluid method. Int J Res Appl Sci Eng Technol 10:5015–5020
9. Odziomek MJ (2017) Colloidal synthesis and controlled 2D/3D assemblies of oxide nanoparticles. Doctoral dissertation, Université de Lyon, AGH University of Science and Technology (Cracovie, Pologne)
10. Taguchi M, Takami S, Adschiri T, Nakane T, Sato K, Naka T (2012) Synthesis of surface-modified monoclinic ZrO_2 nanoparticles using supercritical water. CrystEngComm 14(6):2132–2138

11. Lamastra FR, Bianco A, Meriggi A, Montesperelli G, Nanni F, Gusmano G (2008) Nanohybrid PVA/ZrO$_2$ and PVA/Al$_2$O$_3$ electrospun mats. Chem Eng J 145(1):169–175
12. Tan Y, Sha L, Qu J, Jiang J, Ren J, Wu C, Xu Z (2021) Oleic acid as grinding aid and surface antioxidant for ultrafine zirconium hydride particle preparation. Appl Surf Sci 535:147688
13. He X, Wang Z, Wang D, Yang F, Tang R, Wang JX, Pu Y, Chen JF (2019) Sub-kilogram-scale synthesis of highly dispersible zirconia nanoparticles for hybrid optical resins. Appl Surf Sci 491:505–5168
14. 高淑雅 (2017) Development of high-performance ceramic nanofibers using electrospinning
15. Guerbous L. Structural and optical properties of PVA/ZrO$_2$: Eu^{3+} hybrid films prepared via Γ-irradiation
16. Krzak J, Borak B, Łukowiak A, Donesz-Sikorska A, Babiarczuk B, Marycz K, Szczurek A (2016) Advancement of surface by applying a seemingly simple sol-gel oxide materials. In: Advanced surface engineering materials. Wiley, Hoboken, NJ, pp 33–96
17. Ayotte G, Archambault L, Gingras L, Lacroix F, Beddar AS, Beaulieu L (2006) Surface preparation and coupling in plastic scintillator dosimetry. Med Phys 33(9):3519–3525

Chapter 6
Enhancing Efficiency of Plastic Scintillators

Abstract This study explores the optimization strategies to enhance the efficiency of plastic scintillators used in radiation detection. Improvements in material composition, such as the selection of appropriate polymers and scintillating dopants, are emphasized for optimizing light generation and energy resolution. The integration of wavelength shifters and optical filters is proposed to modify emission wavelengths for better detection sensitivity. Efficient scintillator-photodetector interfacing and the use of nanoparticles like zirconium dioxide (ZrO_2) are discussed to improve photon yield, energy resolution, and overall detection efficiency. Experimental results demonstrate a significant enhancement in gamma detection efficiency with the integration of ZrO_2 nanomaterials, validating the effectiveness of these optimization strategies.

6.1 Introduction

Improving the material composition, including the selection of scintillating polymers and additives to optimize light generation and energy resolution, is necessary to increase the efficiency of plastic scintillators. By using optical filters or wavelength shifters, light emissions may be changed to wavelengths that are more suited for detection. Precise scintillator geometry construction and efficient interfacing with photodetectors, including silicon photomultipliers or photomultiplier tubes, are necessary to increase light collecting efficiency. Optimizing for external variables such as temperature and moisture swings, as well as reducing aging and radiation damage, improves efficiency even more. Strict quality control throughout manufacture guarantees reliable, high-caliber scintillators for precise radiation detection [1].

Improving the effectiveness of plastic scintillators requires a multimodal strategy that takes into account several variables that have a big impact on how well they detect radiation. The selection of base polymers, such as polystyrene or polyvinyl toluene, along with the right scintillating dopants, directly affects the amount of light produced and the energy efficiency of the scintillator. This makes material composition fundamental [2]. Usually, substances like anthracene, stilbene, or other

© The Author(s), under exclusive license to Springer Nature Switzerland AG 2024
R. Naik et al., *Advances in Space Radiation Detection*,
SpringerBriefs in Molecular Science, https://doi.org/10.1007/978-3-031-74551-5_6

organic substances that emit light when activated by ionizing radiation are doped into plastic scintillators. The dopant's properties, including concentration and capacity to effectively transform input radiation energy into visible light, determine the way this light is produced [3].

A further crucial element is wavelength shifting, scintillation light is frequently released at wavelengths that are not optimal for photomultipliers or other light sensors to detect. To get around this, materials that shift wavelengths or optic filters are employed to change the light that is released into a range that corresponds to the maximal sensitivity of the detecting apparatus. The total effectiveness of light gathering and measurement is increased by this method. Another important factor is light collection efficiency, which has to do with how well the scintillation light is captured and transferred to the photodetector. The scintillator's size, shape, and other geometric characteristics affect how successfully light is gathered and transmitted [4].

It is necessary to adjust the scintillator-photodetector connection to guarantee optimal light transmission and low loss, for instance using a silicon photomultiplier (SiPM) or a photomultiplier tube (PMT). To improve the transmission of light and gathering, suitable optical coupling materials and methods are used, such as reflecting coatings or optical gels. Plastic scintillators' performance can also be affected by external variables like temperature and humidity [5]. The luminescence and resolution of energy of many scintillators can change due to temperature variations. Temperature increases, for instance, can shorten the scintillator's decay period and decrease its light production. When the scintillating material is not adequately sealed or protected, humidity might cause it to deteriorate. Other variables influencing efficiency over time include aging and radiation damage [6].

To improve performance, scintillator design optimization includes adjusting the scintillator's size, shape, and surface treatment. For instance, surface treatments that increase surface reflection or decrease surface defects can greatly increase the efficiency of light gathering. Enhancing light production and quickening response times can also be achieved by using sophisticated polymer blends or outstanding performance dopants in scintillator designs [7]. Plastic scintillators may be made much more efficient by tackling these many aspects through advances in material science, engineering, and manufacturing disciplines. This all-encompassing strategy guarantees that plastic scintillators continue to be efficient and dependable instruments for a variety of detection of radiation applications, ranging from particle physics to healthcare imaging as well as beyond [8].

6.2 Calculation Methods for Assessing the Efficiency of Radiation Detectors

Radiation detector effectiveness is determined using several crucial techniques to precisely evaluate the device's performance. The ratio of the total amount of radiation incidents impinging on the detector to the number of detected radiation incidents is used to calculate absolute efficiency. This is usually computed with a calibration source whose radiation output is known so that the reaction time of the detector is precisely measured and compared. By comparing a detector's performance to that of a standard or reference detector, relative efficiency sheds light on how effective a detector is in comparison to pre-established standards. Using conventional sources of radiation along with contrasting the detector's response to a recognized, high-efficiency detector are common procedures for this technology [9].

Another crucial statistic is Energy Resolution, which is determined by examining the detector's capacity to differentiate between various radiation energies. To accomplish this, the full width at half maximum (FWHM) of the energy peaks in the detector's spectrum must be measured and compared to the radiation source's known energy. The minimal detectable activity (MDA), which is frequently determined by statistical analysis, is the lowest quantity of radiation that the detector can consistently detect to evaluate detection sensitivity. To provide accurate measurements over a range of radiation kinds and intensities, calibration factors are computed to modify the raw detector results depending on known radiation sources. With the use of these techniques, radiation detectors' efficiency and efficacy may be better understood, leading to accurate evaluations and advancements in both application and construction [10, 11].

6.3 Experimental Results and Analysis of Efficiency Improvements with Integrated ZrO_2 Nanomaterials

Zinc dioxide (ZrO_2) and other compounds can be added to plastic scintillators to increase their detection effectiveness by improving photon yield and energy resolution. By raising the scintillator's refractive index, nanoparticles improve light collecting and lower scattering losses. This leads to improved resilience, improved energy resolution, and a 30% increase in light output, all of which improve radiation detection accuracy [12].

6.4   Preparation of Nanoparticle Loaded Plastic Scintillators

ZrO_2 nanoparticles were incorporated into standard plastic scintillators such as BC-490 from Saint Gobain. BC-490 is a vinyl-toluene solvent-based partially-polymerized formulation for plastic scintillators, was used as the host matrix of conventional plastic scintillator. The higher viscosity of BC-490 enabled the suspension of nanoparticles inside the plastic scintillator without any sedimentation of the particles, even with 14-day long polymerization. The method outlined in the datasheet of BC-490 was followed for the realization of unloaded plastic scintillators. All reagents were used as received. A proportionate mixture of catalyst and solvent was mixed. It was added to the resin proportionately. Vacuum settling at 10 mbar was performed for 10 min to get rid of air bubbles. The mixture was cast into glass test-tube molds of inner diameter about 9 mm and height of 12 mm. Vacuum settling at 10 mbar was performed for 10 min, again. Molds were kept immersed in de-ionized water maintained at 47 °C for 14 days, inside an oven after which the water was drained, and molds were transferred to the oven for storage in a nitrogen atmosphere at 80 °C for 8 h. The molds were taken out of the oven, and glass was broken in running water. For the realization of nanocomposites, the nanoparticles were added to the resin after the first vacuum settling in the above sequence. Nanoparticles were added directly by weight. Loading was limited to 2 wt % to avoid ambiguous results due to quenching. The mixture was thoroughly stirred. Three samples for each variation were realized. Nanocomposites thus realized were cut to 10 mm thickness. Both faces were polished using a micro grinder with down to 0.05 μm sized alumina powder.

6.5   Measurement of Gamma-Detection Efficiency

Enhancement of gamma detection efficiency was measured by obtaining pulse height spectra for gamma-rays from ^{241}Am for unloaded plastic scintillators and nanocomposites. It was carried out at Space Astronomy Lab at URSC. The set-up had a light-tight enclosure inside which a photomultiplier tube (PMT) (ET Enterprises 9078B) was positioned vertically up. Scintillators were placed over the head of the PMT. They were optically coupled to the head with Dow-corning DC 3500 grease. The enclosure was closed with an aluminum cover with its central portion optically closed with black polythene sheets. The gamma-source was positioned over the aluminium cover. The output of PMT was fed to a multichannel analyzer (MCA, Amptek 8000A) through a charge-sensitive pre-amplifier cum shaping amplifier (Amptek A-203). Data were acquired for 60 s. Background counts were as low as 2 cps against ~ 2500 cps for unloaded plastic scintillator. Under identical measurement conditions, pulse height

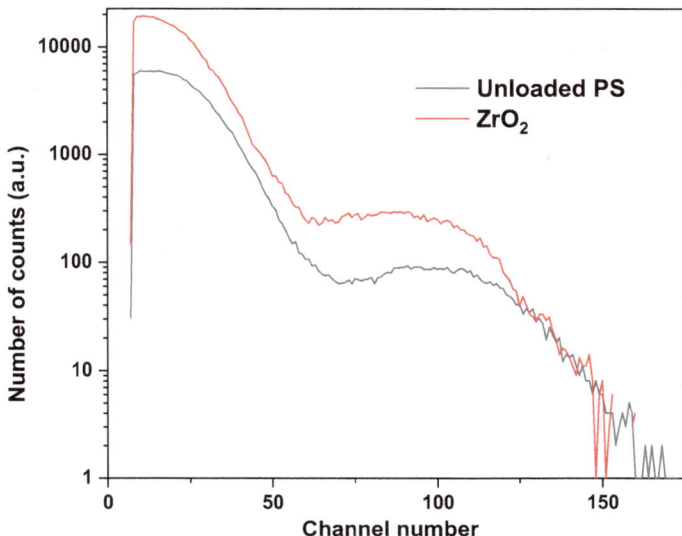

Fig. 6.1 Pulse height spectra of unloaded plastic scintillator realized out of BC-490, and ZrO_2 nanocomposites for gamma-rays from [241]Am, under identical conditions. Channel numbers are curtailed as the number of counts beyond channel 175 were negligible

spectrum for each of the samples was obtained. The total number of counts in the pulse height spectrum indicated a measure of photons emitted by the nanocomposite. Since the radiation input remained the same, the variation in total counts suggested relative gamma detection efficiencies of the nanocomposites [13–15] (Fig. 6.1).

As observed in Fig. 6.1, scintillation with peak energy occurs at around channel 100. The number of counts at this channel for ZrO_2 nanocomposites are greater than those for unloaded plastic scintillator. An enhancement of detection efficiency of 2.7 times is observed for 1 wt % loading of ZrO_2 nanoparticles. The nanoparticles embedded in the scintillator interacted with gamma-rays to emit photoelectrons. The emitted photoelectrons caused additional scintillation in the medium to enhance the counts. Thus, the enhancement of detection efficiency of plastic scintillators upon loading the nanoparticles is demonstrated [16, 17].

6.6 Conclusion

The efficiency of plastic scintillators can be significantly enhanced through a combination of material optimization, geometric refinement, and nanoparticle integration. The study demonstrates that the incorporation of ZrO_2 nanoparticles improves light collection, reduces scattering losses, and increases photon yield, resulting in a 2.7 times enhancement in gamma detection efficiency. These findings highlight the potential for advanced material and engineering strategies to improve the performance of plastic scintillators in various radiation detection applications, including healthcare imaging and particle physics.

References

1. Hajagos TJ, Liu C, Cherepy NJ, Pei Q (2018) High-Z sensitized plastic scintillators: a review. Adv Mater 30(27):1706956
2. Wang Z, Dujardin C, Freeman MS, Gehring AE, Hunter JF, Lecoq P, Liu W, Melcher CL, Morris CL, Nikl M, Pilania G, Pokharel R, Robertson DG, Rutstrom DJ, Sjue SK, Tremsin AS, Watson SA, Wiggins BW, Winch NM, Zhuravleva M (2023) Needs, trends, and advances in scintillators for radiographic imaging and tomography. IEEE Trans Nucl Sci 70(7):1244–1280
3. Bingham S, Daoud WA (2011) Recent advances in making nano-sized TiO_2 visible-light active through rare-earth metal doping. J Mater Chem 21(7):2041–2050
4. Yanagida T (2018) Inorganic scintillating materials and scintillation detectors. Proc Jpn Acad Ser B 94(2):75–97
5. Dahlbom M (2017) Photodetectors. In: Physics of PET and SPECT imaging. CRC Press, pp 85–112
6. Mikhailik VB, Kraus H (2010) Performance of scintillation materials at cryogenic temperatures. Phys Stat Sol (b) 247(7):1583–1599
7. El-Mohri Y, Antonuk LE, Choroszucha RB, Zhao Q, Jiang H, Liu L (2014) Optimization of the performance of segmented scintillators for radiotherapy imaging through novel binning techniques. Phys Med Biol 59(4):797
8. Wibowo A, Sheikh MAK, Diguna LJ, Ananda MB, Marsudi MA, Arramel A, Zeng S, Wong LJ, Birowosuto MD (2023) Development and challenges in perovskite scintillators for high-resolution imaging and timing applications. Commun Mater 4(1):21
9. Hassler DM, Zeitlin C, Wimmer-Schweingruber RF, Böttcher S, Martin C, Andrews J, Böhm E, Brinza DE, Bullock MA, Burmeister S, Ehresmann B, Epperly M, Grinspoon D, Köhler J, Kortmann O, Neal K, Peterson J, Posner A, Rafkin S, Seimetz L, Smith KD, Tyler Y, Weigle G, Reitz G, Cucinotta FA (2012) The radiation assessment detector (RAD) investigation. Space Sci Rev 170:503–558
10. Statham PJ (1976) A comparative study of techniques for quantitative analysis of the X-ray spectra obtained with a Si (Li) detector. X-Ray Spectrom 5(1):16–28
11. Done L, Ioan MR (2016) Minimum detectable activity in gamma spectrometry and its use in low level activity measurements. Appl Radiat Isot 114:28–32
12. Rogenski EN (2021) The investigation of AM ceramics for the production of a 3D printed high temperature thermocouple. Youngstown State University
13. Renuka L, Anantharaju KS, Sharma SC, Nagabhushana H, Vidya YS, Nagaswarupa HP, Prashantha SC (2017) A comparative study on the structural, optical, electrochemical and photocatalytic properties of ZrO_2 nanooxide synthesized by different routes. J Alloy Compd 695:382–395
14. Pratapkumar C, Prashantha SC, Nagabhushana H, Anilkumar MR, Ravikumar CR, Nagaswarupa HP, Jnaneshwara DM (2017) White light emitting magnesium aluminate nanophosphor: near ultra violet excited photoluminescence, photometric characteristics and its UV photocatalytic activity. J Alloy Compd 728:1124–1138
15. Gurushantha K, Renuka L, Anantharaju KS, Vidya YS, Nagaswarupa HP, Prashantha SC, Nagabhushana H (2017) Photocatalytic and photoluminescence studies of ZrO_2/ZnO nanocomposite for LED and waste water treatment applications. Mater Today Proc 4(11):11747–11755
16. Gurushantha K, Anantharaju KS, Nagabhushana H, Sharma SC, Vidya YS, Shivakumara C, Nagaswarupa HP, Prashantha SC, Anilkumar MR (2015) Facile green fabrication of

iron-doped cubic ZrO_2 nanoparticles by *Phyllanthus acidus*: structural, photocatalytic and photoluminescent properties. J Mol Catal A Chem 397:36–47

17. Vidya YS, Anantharaju KS, Nagabhushana H, Sharma SC, Nagaswarupa HP, Prashantha SC, Shivakumara C (2015) Combustion synthesized tetragonal ZrO_2: Eu^{3+} nanophosphors: structural and photoluminescence studies. Spectrochim Acta Part A Mol Biomol Spectrosc 135:241–251

Chapter 7
Applications of Advanced Radiation Detection Systems

Abstract This chapter explores the diverse applications of advanced radiation detection systems across multiple fields. These systems, incorporating technologies such as scintillation detectors, high-purity germanium detectors, and innovative data processing methods, significantly enhance radiation detection and quantification capabilities. In space exploration, they monitor cosmic and solar radiation to protect astronauts and equipment. In nuclear energy, they ensure safety in power plants and waste management. In medical fields, they improve diagnostic imaging and radiation therapy accuracy. These technologies also play critical roles in environmental monitoring, climate studies, wildlife protection, homeland security, and industrial processes. As advancements continue, including integration with AI and IoT, radiation detection systems will become more effective, supporting safety, regulatory compliance, and innovative research across a wide range of domains.

7.1 Introduction

Significant advancements have been made in advanced radiation detection systems, improving the capacity to precisely detect and quantify radiation. Among the latest developments are scintillation detectors, which transform radiation into light signals by using substances like sodium iodide [1]. For gamma-ray spectroscopy, high-purity germanium detectors offer remarkable resolution. More precision and stability are provided by sophisticated ionization chambers. New methods are emerging for data processing, such as dosimetry which employs artificial intelligence and nanotechnology. The combined effect of these developments enhances detection sensitivity and application across a range of domains, ranging from space exploration to nuclear safety [2].

Advanced radiation detection systems have significance in a variety of sectors because of their capacity to improve security and efficiency. They enhance imaging for diagnosis and treatment of cancer in healthcare by monitoring radiation doses

R. Naik et al., *Advances in Space Radiation Detection*,
SpringerBriefs in Molecular Science, https://doi.org/10.1007/978-3-031-74551-5_7

precisely. These technologies improve safety in nuclear power and waste management by detecting and measuring radioactivity [3]. They additionally have an important role in environmental monitoring, measuring pollution levels, and safeguarding ecosystems. In space missions, they protect astronauts from cosmic rays by real-time monitoring. These mechanisms are critical to safety, research, and compliance with regulation [4]. The objective of this chapter is to examine the wide range of uses of contemporary radiation detection technology in different fields. It will address their application in nuclear power plants and waste management for safety and compliance with regulations, in healthcare for accurate radiation therapy and imaging, and in environmental monitoring to identify and evaluate radioactive contamination. The chapter will also discuss their use in space exploration to safeguard humans and monitor cosmic radiation. Every application shows how important these systems are to research, safety, and efficiency in operation.

7.2   Space Exploration

7.2.1   Space Radiation Monitoring

Radiation detection devices are critical for tracking different space radiation circumstances in space exploration. These systems monitor solar radiation, notably solar particle events (SPEs), which may have a major impact on spacecraft and astronauts, in addition to cosmic rays, which are particles with high energies that come from beyond the solar system. Particle flux and energy spectrum are measured by sophisticated detectors, which include scintillation counters and silicon detectors, which provide vital information for mission preparation and shielding tactics. In addition, space radiation sensors are essential for both mission success and astronaut safety since they aid in evaluating radiation exposure during extended missions. These systems facilitate the research and creation of defences against radiation-induced damage by assisting in understanding the effects of radiation on biological structures [5].

7.2.2   Human Spaceflight Safety

Innovative radiation detection systems monitor and reduce hazardous radiation exposure, which is essential for the protection of astronauts on missions to space. These devices quantify the amounts of ionizing radiation that comes from cosmic rays and SPEs, which include dosimeters and real-time radiation monitors [4]. They aid in estimating radiation dose rates and forecasting high-radiation occurrences by offering continuous data, which is essential for prompt protective action. Modern technologies guarantee that an astronaut's exposure stays within acceptable ranges and allow for timely modifications to mission plans. Technologies include individual dosimeters and the Radiation Assessment Detector (RAD) aboard Mars rovers. The establishment of efficient radiation shielding plans and responses is facilitated by these detection systems, which improves crew security and the achievement of the mission [6].

7.2.3 Spacecraft and Satellite Protection

Modern radiation detection techniques are crucial for shielding electronic parts from damage caused by radiation in satellites and spacecraft. Radiation sensors in these systems keep an eye out for cosmic rays and high-energy particles that might harm spacecraft electronics and cause malfunctions or deterioration. Spacecraft may reduce the impact of radiation on fragile electronics by combining real-time radiation sensors with shielding techniques including irradiation-hardened materials and proactive shields. By preserving the lifetime and performance of onboard equipment, this proactive strategy contributes to mission success and dependability [7].

7.3 Nuclear Energy and Safety

7.3.1 Nuclear Power Plants

To maintain acceptable radiation levels, these systems incorporate dosimeters and real-time radiation monitors that track radiation fields inside reactor cores and the surrounding surroundings. To prevent mishaps, they assist in leak detection, radioactive waste management, and plant status monitoring. These technologies increase the efficiency of plants, assist with compliance with regulations, and improve worker safety by continually evaluating radiation exposure [8].

7.3.2 Nuclear Waste Management

Nuclear waste must be handled, stored, and disposed of safely, which requires the use of sophisticated radiation detectors. These devices, which track radiation levels and discover radioactive isotopes in the trash, encompass high-resolution gamma spectrometers and neutron detectors. Offering real-time data on radiation emissions and aiding in the management of waste storage facilities, they guarantee precise waste classification, safe containment, and appropriate disposal. During the waste management process, efficient usage of these detectors reduces the negative effects on the environment and guarantees compliance with regulations [9].

7.3.3 Radiation Emergency Response

Advanced detection systems are critical for controlling radiation leakages and disasters in radiation response to emergencies. These technologies offer real-time information on radiation levels and contamination spread during accidents. These systems include portable radiation detectors and aerial monitoring devices. Expedite the process of evaluating affected regions, allowing for prompt evacuation and purification actions. Effective emergency preparedness and response strategies are ensured by these detectors, which assist in locating radiation sources and evaluating public health dangers. When they are deployed safety in general is improved and the effects of radiation crises are lessened [10].

7.4 Medical Applications

7.4.1 Diagnostic Imaging

By increasing picture quality and diagnostic precision, advanced radiation detection systems improve medical imaging procedures like CT, PET, and X-ray scans. Excellent quality detectors and digital imaging sensors enhance brightness and spatial accuracy in X-ray and CT images, which helps in early illness diagnosis. PET scans provide precise monitoring of the metabolism by measuring gamma rays with great accuracy using sophisticated scintillation detectors. These technologies improve safety for patients and diagnostic capacities by lowering radiation levels while offering comprehensive anatomic and functional data [11].

7.4.2 Radiation Therapy

Modern detecting technologies are essential for accurately administering doses of radiation therapy and enhancing the results of cancer treatments. Ionization chambers and dosimeters are examples of systems that guarantee precise dosage measurement and distribution while reducing harm to nearby healthy tissue. Accurate tumor targeting and dosage verification are made possible by real-time imaging technologies, such as in-line CT and MRI. These developments support the optimization of therapeutic efficacy, the reduction of adverse effects, and the dynamic adaptation of treatment strategies. Enhanced patient safety and customized treatment strategies are greatly aided by enhanced monitoring techniques [12].

7.4.3 Radiopharmaceuticals

Modern radiation detection systems are essential to therapeutics and diagnostics processes in radioactive substances. Gamma cameras and PET scanners are instances of diagnostic technologies that use radiation detection from radiopharmaceuticals to see physiological processes and provide very accurate diagnoses. To guarantee precise administration and evaluate the efficacy of treatment, these detectors track the radiation released by therapeutic radiopharmaceuticals. Utilizing imaging technology and excellent-quality detectors improves dosage administration accuracy and aids in therapy result optimization [13].

7.5 Environmental Monitoring

7.5.1 Radiation in the Environment

Monitoring ambient levels of radiation resulting from natural and man-made sources requires sophisticated detection devices. These devices offer immediate information on gamma rays, beta particles, and radiation levels. They include scintillation counters, ambient radiation sensors, and compact detectors. They provide the tracking of radiation from both natural and anthropogenic sources, including nuclear power plants and radioactive contamination. Natural sources of radiation include cosmic radiation and indigenous radionuclides. These devices promote public health safety, safeguarding the environment, and regulatory compliance by providing comprehensive geographical and temporal exposure patterns [14].

7.5.2 Climate Change Studies

Scientific research on the effects of radiation on meteorological and climate-related events requires sophisticated radiation detection devices. These systems track whether cosmic rays impact the chemistry of the atmosphere, depletion of ozone, cloud formation, and climate trends. Researchers can determine whether radiation influences climatic variables like precipitation and temperature by examining information regarding radiation levels and their interactions with atmospheric components. The function of irradiation in climate change is better understood owing to this research, which also aids in the creation of prediction models for ecological and climate impact evaluations [15].

7.5.3 Wildlife and Ecosystem Monitoring

In regions where nuclear activity has taken place, sophisticated radiation detection devices are essential for tracking radiation levels that damage ecosystems and animals. These devices are used to monitor radiation in ecosystems and evaluate its impacts on vegetation and wildlife, and they include biological tests and field irradiation sensors. Researchers assess the effects of radiation on biotic health, success in reproduction, and ecosystem stability by monitoring levels of contamination and biological absorption of radionuclides [16]. By leading remediation activities to safeguard impacted ecosystems, this monitoring aids in evaluating long-term ecological impacts.

7.6 Homeland Security and Defense

7.6.1 Border and Port Security

Sophisticated radiation detection devices are essential for border and port safety since they stop radioactive goods from being smuggled. These devices, which include portable spectroscopy instruments neutron sensors, and gamma-ray detection devices, are used to search people, cars, and cargo for radioactive materials that are not permitted. Materials that may be utilized in hazardous bombs or other radioactive dangers are identified and intercepted with their assistance [17]. Authorities may better identify and stop the illegal trafficking of radioactive materials by incorporating these tools into security procedures, which protects both national security and the general public.

7.6.2 Nuclear Non-proliferation

In the framework of nuclear non-proliferation, sophisticated radiation detection devices are essential for tracking adherence to global accords and identifying illicit nuclear operations. These tools are employed to find and identify unlawful nuclear substances and activities. They include spectrometers, environmental sampling devices, and remote sensing technologies. They offer essential data for confirming nuclear disarmament and keeping surveillance on nuclear installations to make sure they follow non-proliferation agreements. Efficient detection aids in the implementation of international laws and stops the proliferation of weapons of mass destruction and components [18].

7.6.3 Counterterrorism

In the fight against terrorism, advanced radioactivity detection devices are essential for recognizing and preventing the possession of "dirty explosives", or dangerous weapons. These devices are designed to detect and evaluate radioactive elements in a variety of settings. They include aerial sensors, portal monitors, and portable radiation detectors. They aid in determining the radioactive content of suspected gadgets or cargo and offer early warning of possible risks. Authorities can improve their capacity to stop radioactive assaults and safeguard public safety by incorporating these advances into security procedures [19].

7.7 Industrial Applications

7.7.1 Quality Control in Manufacturing

Non-destructive testing (NDT) for inspection of materials and assurance of quality during production requires radiation detection equipment as a fundamental component. X-rays and gamma rays are used in radiography testing procedures to identify internal flaws and verify the structural strength of components without causing harm to individuals. According to these advances offer precise pictures that aid in locating flaws in metals and other materials, such as fractures or inconsistent welding. Radioactive detection improves dependability and adherence to industrial requirements by guaranteeing the caliber and security of produced products [20].

7.7.2 Mining and Resource Exploration

In mining and resource development, radiation detectors are critical for finding and evaluating radioactive materials. Radioactive elements in mineral deposits are identified and measured using gamma-ray spectrometers, scintillation detectors, and Geiger counters. These mechanisms aid in determining the amount of precious minerals like thorium and uranium, directing the extraction process, and guaranteeing effective resource management. These devices help ensure safety in the environment and regulatory compliance over mining operations by giving real-time data on the amount of radiation [21].

7.7.3   Radiation-Based Process Monitoring

Radiation-based process monitoring is critical in sectors that use sterilization and material manipulation. Radioactive detectors make certain that gamma rays or electron beams are delivered precisely during sterilization so that bacteria are destroyed efficiently and products are not harmed. To manage polymer cross-linking and other treatments that improve material characteristics, technologies such as radiographic sensors track radiation exposure. These detecting systems offer real-time feedback, guaranteeing process uniformity, safety, and quality of the product [22].

7.8   Research and Development

7.8.1   Fundamental Research

Advanced radiation detection devices play a crucial part in fundamental studies in nuclear physics and astronomy. These devices, which include detectors for particles and excellent-quality gamma-ray spectrometers, allow for the precise evaluation of the reactions and decay processes in the field of nuclear physics. The research and investigation of cosmological radiation, mysterious matter, and high-energies astrophysical events are made easier in astrophysics by monitors such as Cherenkov telescopes and space-based gamma-ray observatories. These technologies advance mathematical models and experimental procedures by providing vital data for comprehending basic particles and cosmic events [23].

7.8.2   Innovation in Detector Technology

The development of novel materials, technologies, and procedures is greatly aided by advancements in detection technology. To enhance radiation detection, new scintillator materials with increased efficiency and resolution have been developed recently. Higher sensitivities and resolution of energy for accurate measurements are provided by advancements in solid-state detectors, such as high-purity germanium and diamond sensors. Furthermore, new approaches include artificial intelligence algorithms to improve data analysis and interpretation, thereby boosting the effectiveness and capability of identification. These developments propel advancements in a multitude of domains, ranging from monitoring the environment to imaging for medical purposes [24].

7.8.3 Interdisciplinary Applications

Advanced radiation detection systems are becoming more prevalent in multidisciplinary study fields that involve medicine and ecology. In biological processes, these devices are used to follow radiolabelled tracers in cellular and biochemical investigations, which helps researchers comprehend the causes of illness and interactions between therapies. Environmental scientists monitor radioactive pollutants in soil, water, and air to determine the effects of radiation on habitats and the general population. These applications improve research capacities across disciplines by delivering accurate and real-time data, resulting in advances regarding both scientific knowledge and real-world solutions [25].

7.9 Future Perspectives and Emerging Applications

7.9.1 Innovations on the Horizon

Potential uses of enhanced radiation detection systems are expected to influence various new sectors. Customized nano detectors, for instance, have the potential to improve medical diagnostics by enabling more accurate and less intrusive imaging approaches. Advances in ultra-sensitive detectors have the potential to improve cosmic ray detection and enable distant flights. Furthermore, new technologies such as quantum sensors may provide unparalleled accuracy and precision for radiation detection, allowing advances in basic physics and material science to continue. These innovations have the potential to accelerate progress in a variety of scientific and economic domains [14].

7.9.2 Integration with Other Technologies

The capabilities of radiation detection systems will be improved by integrating them with AI, IoT, and other cutting-edge technology. AI systems evaluate radiation detector data in real-time, which enhances anomaly identification, predictive maintenance, and threat detection. Real-time environment and security evaluations are made possible by IoT integration, which enables smooth data transfer and remote monitoring. Furthermore, integrating blockchain technology with these systems may provide safe and impenetrable data management. By offering more precise and rapid solutions, these linkages hold the potential to enhance applications in environmental monitoring, security, and health [26].

7.9.3 Global Impact and Collaboration

Technologies for radiation detection have a significant worldwide influence on everything from national security and environmental preservation to public health and safety. For these innovations to advance, satisfy global requirements, and solve cross-border issues, international collaboration is essential. Through cooperative research projects and data sharing, collaborative activities improve information sharing, standardize procedures, and promote innovation. This kind of collaboration makes it easier to respond to global concerns like environmental degradation and nuclear proliferation, highlighting the necessity of a coordinated strategy to address these intricate, worldwide problems [18].

7.10 Conclusion

Sophisticated radiation detection systems can be helpful in many different sectors, improving both effectiveness and security. To safeguard personnel and spaceships, they measure solar and cosmic radiation throughout space missions. They oversee nuclear waste management as well as security at power plants in the field of nuclear energy. Radiation therapy and diagnostic imaging are two medical applications that increase therapeutic accuracy. These systems are used in environmental monitoring to monitor radioactive pollution and evaluate its influence on the environment. These numerous applications highlight how important irradiation detection technology is to solving today's problems.

Sophisticated radiation detection systems are essential for many companies and scientific domains, providing valuable information and improving safety. While they control waste and preserve the security of operations in nuclear power, their position in space exploration guarantees astronaut protection and spacecraft integrity. These systems enhance the effectiveness of treatment and diagnostic precision in medicine. Precise radiation measurements assist process efficiency and public health in monitoring the environment and applications in industry. To solve complicated problems and spur innovation, these innovations must be continuously improved upon and integrated.

Radiation detection system application, research, and development must continue to successfully address worldwide concerns. Sustained progress can boost safety measures, increase detection precision, and propel technical innovations in domains including monitoring the environment, nuclear energy, and the exploration of space. Scientists, business executives, and legislators working together will hasten the development of these technologies and guarantee that they satisfy new demands. Stronger answers to intricate global problems will be made possible by funding state-of-the-art research and encouraging multidisciplinary collaborations, underscoring the vital role radiation detection devices play in preserving the future of humanity.

References

1. Knoll GF (2010) Radiation detection and measurement. Wiley
2. Lagopati N (2021) Nanotechnology in nuclear medicine/MATLAB use. In: Clinical nuclear medicine physics with MATLAB®. CRC Press, pp 325–338
3. Kakavelakis G, Gedda M, Panagiotopoulos A, Kymakis E, Anthopoulos TD, Petridis K (2020) Metal halide perovskites for high-energy radiation detection. Adv Sci 7(22):2002098
4. Montesinos CA, Khalid R, Cristea O, Greenberger JS, Epperly MW, Lemon JA, Boreham DR, Popov D, Gorthi G, Ramkumar N, Jones JA (2021) Space radiation protection countermeasures in microgravity and planetary exploration. Life 11(8):829
5. Hu S (2017) Solar particle events and radiation exposure in space. NASA Space Radiation Program Element, Human Research Program, pp 1–15
6. Vangen SD (2011) NASA technology development entries for ISECG TAT (Technology Assessment Team). In: ISECG working group's workshop (No. JSC-CN-23747)
7. Rogers SC (1963) Radiation damage to satellite electronic systems. IEEE Trans Nucl Sci 10(1):97–105
8. Adamantiades A, Kessides I (2009) Nuclear power for sustainable development: current status and future prospects. Energy Policy 37(12):5149–5166
9. McManus KD (2019) Radiological and nuclear threat detection using small unmanned aerial systems. University of California, Berkeley
10. Ahmad MI, Rahim MH, Nordin R, Mohamed F, Abu-Samah A, Abdullah NF (2021) Ionizing radiation monitoring technology at the verge of internet of things. Sensors 21(22):7629
11. Hussain S, Mubeen I, Ullah N, Shah SSUD, Khan BA, Zahoor M, Ullah R, Khan FA, Sultan MA (2022) Modern diagnostic imaging technique applications and risk factors in the medical field: a review. BioMed Res Int 2022(1):5164970
12. Abshire D, Lang MK (2018) The evolution of radiation therapy in treating cancer. Semin Oncol Nurs 34(2):151–157
13. Pathak A (2023) Use of radiation in diagnosis. In: Tools and techniques in radiation biophysics. Springer Nature, Singapore, pp 163–175
14. Marques L, Vale A, Vaz P (2021) State-of-the-art mobile radiation detection systems for different scenarios. Sensors 21(4):1051
15. Kirkby J (2007) Cosmic rays and climate. Surv Geophys 28:333–375
16. Geras'kin SA (2016) Ecological effects of exposure to enhanced levels of ionizing radiation. J Environ Radioact 162:347–357
17. Kouzes R (2021) Radiation detection technology for homeland security. In: Handbook of particle detection and imaging. Springer International Publishing, Cham, pp 897–927
18. Zubair M, Radkiany R, Akram Y, Ahmed E (2024) Nuclear safeguards: technology, challenges, and future perspectives. Alex Eng J 108:188–205
19. Kumar A, McCue J, Evans A (2024) Physical protection of nuclear facilities and materials. In: The challenges of nuclear security: US and Indian perspectives, p 159
20. Gupta M, Khan MA, Butola R, Singari RM (2022) Advances in applications of Non-Destructive Testing (NDT): a review. Adv Mater Process Technol 8(2):2286–2307
21. Kiprono NR, Smolinski T, Rogowski M, Chmielewski AG (2023) The state of critical and strategic metals recovery and the role of nuclear techniques in the separation technologies development. Separations 10(2):112
22. Singh S, Mehta D (2020) Sterilization of pharmaceutical dosage forms. Expectations and realities of multifunctional drug delivery systems. Drug Deliv Asp 4:169
23. Wang Z, Dujardin C, Freeman MS, Gehring AE, Hunter JF, Lecoq P, Liu W, Melcher CL, Morris CL, Nikl M, Pilania G, Pokharel R, Robertson DG, Rutstrom DJ, Sjue SK, Tremsin AS, Watson SA, Wiggins BW, Winch NM, Zhuravleva M (2023) Needs, trends, and advances in scintillators for radiographic imaging and tomography. IEEE Trans Nucl Sci 70(7):1244–1280
24. Lin Z, Lv S, Yang Z, Qiu J, Zhou S (2022) Structured scintillators for efficient radiation detection. Adv Sci 9(2):2102439

25. Gupta T (2013) Radiation, ionization, and detection in nuclear medicine. Springer, Heidelberg, pp 451–494
26. Lu ZX, Qian P, Bi D, Ye ZW, He X, Zhao YH, Su L, Li SL, Zhu ZL (2021) Application of AI and IoT in clinical medicine: summary and challenges. Curr Med Sci 41(6):1134–1150

Chapter 8
Conclusion and Future Perspectives for Space Radiation Detection

Abstract This chapter provides a comprehensive overview of space radiation detection, focusing on technologies used to monitor cosmic and solar radiation to ensure astronaut safety and mission success. It highlights current technologies, such as scintillation detectors, semiconductor detectors, and passive dosimeters, along with their strengths, limitations, and applications in various space missions. The chapter discusses challenges in detecting space radiation, including technical, environmental, and operational difficulties, and outlines emerging technologies like nanomaterial-based detectors, microelectromechanical systems (MEMS), and AI-enhanced data processing. The future of space radiation detection emphasizes integrating advanced sensors with spacecraft systems, improving prediction and response capabilities, and ensuring long-term mission safety. Ongoing innovation and collaboration are crucial to overcoming the complexities of space radiation and advancing human exploration further into space.

8.1 Introduction

8.1.1 Overview of Space Radiation Detection

Space radiation detection entails tracking cosmic waves, solar fragments, and various other irradiation types in space to assure astronaut security and the achievement of their missions. Key topics include galactic cosmic rays (GCRs) and solar energetic particles (SEPs) as well as methods for detecting them that include a solid-state detector, scintillation counters, and dosimeters. These systems use irradiation magnitude, energy, and flux to identify possible dangers and reduce exposure. Effective detection is critical for comprehending how irradiation affects either individuals or spaceship components [1].

© The Author(s), under exclusive license to Springer Nature Switzerland AG 2024
R. Naik et al., *Advances in Space Radiation Detection*,
SpringerBriefs in Molecular Science, https://doi.org/10.1007/978-3-031-74551-5_8

Continuous research and technology breakthroughs in cosmic radiation monitoring are crucial for improving operation reliability and safeguarding personnel from radiation. Steady developments in detecting technology, such as greater sensitivity and precise sensors, aid in the improved tracking of cosmic radiation and solar radiation. Improvements in methods for analysing data, such as algorithms for machine learning, allow for more accurate forecasts and real-time evaluations of radiation dangers [2]. New radiation-resistant fabrics and structures are being researched to reduce radiation. These developments are critical for long-term missions, especially those to Mars as well as beyond, where irradiation poses serious health hazards. Spending and promoting this research assure astronaut's safety and performance in exploring the space missions [3].

The chapter is to give a complete examination of space radiation detection, concentrating on the key components of comprehending and handling radiation in spacecraft settings. The goals are to understand space radiation and seek to clarify the origins and causes of radiation from space, which includes cosmic radiation and coronal intense particles, as well as their possible effects on human health and spaceflight. Evaluating radiation sensing technologies to evaluate existing radiation identification technologies, which include solid-state detectors and dosimeters, and their usefulness in space missions. Addressing technical accomplishments aims to highlight current accomplishments and prospects in radiation detection technology, with a focus on improvements that improve protection and the success of missions [4].

8.2 Summary of Current Space Radiation Detection Technologies

8.2.1 Overview of Technologies

Present cosmic radiation detection methods comprise scintillation detectors, semiconductor detectors, and passive dosimeters. Scintillation sensors employ materials that release light when exposed to radiation, which is subsequently turned into an electrical signal. These are commonly used because of their sensitivity and rapid reaction. Semiconductor detectors, including those made of silicon or germanium, generate charge carriers by the interaction of radiation with a semiconductor material, allowing for great resolution and precision [5]. Passive dosimeters, which include thermoluminescent (TLD) and optically stimulated luminescence (OSL) dosimeters, assess the exposure to radiation by monitoring changes in radiation-absorbing materials as time passes. Every technique has advantages and disadvantages, with scintillation detectors excelling at continuous surveillance, semiconductor detectors offering great precision, and passive dosimeters suited for long-term exposure studies. The type of equipment chosen is determined by the mission's or study's unique needs [6].

A comparison of existing space radiation detection technology reveals various strengths and limits. Scintillation detectors provide great sensitivity and quick response, although they may have low energy precision and thermal susceptibility. Semiconductor detectors, including devices made of silicon, offer outstanding resolution and precision, although they are expensive and prone to harm from radiation over time. Passive dosimeters, which include TLDs and OSLs, are affordable and ideal for ongoing tracking, but they lack real-time measuring capabilities. Each approach has distinct benefits, thus being appropriate for certain parts of space radiation measurement and needing careful consideration based on mission objectives [7].

8.2.2 Case Studies

Table 8.1 shows the comparative analysis of different space mission and radiation detection system. The Radiation Assessment Detector (RAD), which is essential for determining astronaut radiation dosage and directing future human missions, is used by the Mars Science Laboratory (MSL)—Curiosity Rover to track the amount of radiation on the surface of the mars. To continually provide data on irradiation in low Earth orbit and guide shielding and health measures, the International Space Station (ISS) utilizes a variety of detectors, which include those from NASA-LANL and ESA [8]. Measuring cosmic rays and interstellar radiation using the Cosmic Ray Subsystem (CRS), the Voyager Probes improve our knowledge of extremely distant settings [9]. The Lunar Surface Experiments Package was used by the Apollo Missions to collect preliminary data on Moon radiation, which impacted the safety of subsequent lunar missions [10]. Understanding the radiation risks inside planetary magnetospheres was aided by the Heavy Ion Counter (HIC) on the Galileo Mission, which measured radiation in Jupiter's magnetosphere [11].

High radiation levels on Mars, notable variations in radiation intake on the International Space Station, and powerful cosmic rays found outside of the heliosphere are

Table 8.1 Comparative analysis of different space mission and radiation detection system

S. No.	Space mission	Radiation detection system	References
1	Mars Science Laboratory (MSL)—Curiosity Rover	Radiation Assessment Detector (RAD)	[12]
2	International Space Station (ISS)	Various sensors including NASA-LANL and ESA instruments	[8]
3	Voyager Probes	Cosmic Ray Telescope for the Effects of Radiation (CRaTER)	[9]
4	Apollo Missions	Lunar Surface Experiments Package	[10]
5	Galileo Mission	Jupiter Energetic Particle Detector Instrument (JEDI)	[11]

among the key results from case studies on space radiation. These realizations influence the exploration of space by emphasizing the necessity of sophisticated mission planning, surveillance infrastructure, and shielding from radiation. They serve as a roadmap for designing habitats and spaceships for longer, safer journeys.

8.3 Challenges and Limitations

8.3.1 Technical Challenges

Various technological hurdles to detecting space radiation. Hostile space conditions can harm sensitive detectors, resulting in reduced efficiency and calibration concerns. The wide variety of radiation types and energy necessitates detectors with outstanding sensitivity and resolution, which complicates design and data processing. Radiation shielding can be challenging, particularly for long-duration missions, resulting in possible errors. Furthermore, monitoring and interpreting the massive amounts of data produced by radiation detectors presents computing and storage issues. Addressing these concerns necessitates strong, adaptive equipment and improved data handling procedures to provide dependable and precise radiation monitoring in space [2].

Sensitivity, precision, and dependability are common concerns with space radiation detecting systems. Sensitivity is critical to detecting small amounts of radiation, yet devices with great sensitivity might be more prone to interference and noise. Accuracy is required for exact measurements of radiation levels and energy, but attaining it can be difficult owing to calibration errors and other environmental influences. Reliability is critical for maintaining constant performance over time and under changing conditions, yet space settings can degrade equipment or introduce mistakes. To provide effective and reliable space radiation evaluation, detecting systems have to reconcile these aspects [13].

8.3.2 Environmental and Operational Challenges

The space climate has a considerable influence on radiation detector performance. Temperature fluctuations can impact detector materials, producing calibration deviations and sensitivity deterioration. Massive amounts of irradiation which include cosmic radiation and solar particle incidents, can cause detector overload or destruction. Vacuum conditions and micrometeoroid hits increase the danger of physical damage and infection. Furthermore, ongoing exposure to space radiation might result in "radiation hardness" concerns, compromising the detector's long-term dependability and accuracy. Designing detectors to resist these severe circumstances is critical for ensuring accurate and consistent performance during space missions [13, 14].

The radiation from space detectors has operational problems such as regulating energy use, decreasing pounds, and assuring seamless connection with spaceship components. High power consumption might limit the amount of energy accessible to other systems, therefore effective energy utilization is crucial. Weight limits are crucial owing to launching and space flight constraints, consequently, sensors must be both portable and solid. Planetary systems require flawless interaction and information management, as well as interoperability with onboard electronics and software. These operating limits need careful planning and development to guarantee the radiation detectors perform properly without interfering with other mission components. Controlling these elements is critical for successful missions in space [15].

8.4 Future Trends in Space Radiation Detection

8.4.1 Emerging Technologies

High-sensitivity and long-lasting nanomaterial-based detectors, such as those that use graphene or carbon nanotubes, are among the emerging technologies in space radiation detection. Microelectromechanical systems (MEMS) technology has been researched for application in miniature, compact detectors with improved accuracy. Further, quantum dot-based sensors offer higher energy resolution and efficiency. The combination of AI and machine learning improves the analysis of information and pattern identification, resulting in more precise forecasts and continuous surveillance. These advances aim to increase performance, decrease size and weight, and provide greater resistance to hostile space surroundings, hence expanding the possibilities of space radiation detecting systems [16].

Advances in sensing technology, materials science, and information processing have the potential to lead to advances in radiation from space detection. In the field of sensor technology, advances in single-photon detectors and quality image sensors promise remarkable sensitivity and accuracy. Materials research is progressing via the application of innovative radiation-resistant materials and thin-film methods, which improve both performance and durability. Machine learning techniques and quantum computers are being developed to analyse complicated data sets more rapidly, allowing for real-time analysis and better radiation measurement accuracy. These advancements attempt to address present limits by providing more reliable, highly sensitive, and effective methods for detecting space radiation [17].

8.4.2 Integration with Space Missions

Future strategies for the integration of radiation detection devices in space missions emphasize versatility and economy. Advanced sensors will be modularly integrated, enabling quick upgrades and upgrades. Information from detectors that detect radiation will be easily linked with spacecraft systems to enable real-time monitoring and decision-making. Enhanced protocols for communication will guarantee that radiation data is rapidly transmitted to terrestrial control and other satellite systems. Furthermore, integrating AI and machine learning will make it easier to detect radiation dangers and respond to them. These solutions are intended to improve protection against radiation, mission safety, and operational effectiveness in more complex tasks in space [18, 19].

Certain issues will need to be addressed by radiation detection systems for long-duration tasks, deep space discovery, and space travel by humans. It will be essential to have sophisticated, long-lasting detectors to track levels of radiation over time and in a variety of settings. Devices have to be able to precisely identify solar particle events and high-energy cosmic rays to be used for deep space travel. To safeguard astronauts from hazardous radiation, comprehensive radiation protection with real-time health monitoring will be necessary for human spaceflight missions. Additionally, the development of prediction algorithms to foresee the dangers of exposure to radiation and guarantee astronaut safety during extended space missions and trips outside of low Earth orbit will be emphasized [20].

8.5   Conclusion

8.5.1 Summary of Key Finding

Significant aspects of space radiation detection are covered in this chapter, with a focus on the variety of available technologies, such as passive dosimeters, semiconductors, and scintillation. It emphasizes operational problems like energy use, weight, and integration of systems in addition to technical difficulties like sensitivity, accuracy, and dependability. The effects of harsh temperatures and radiation-induced degradation on detector performance are mentioned. Promising developments can be found in emerging technologies, such as sophisticated sensors, unique materials, and AI-powered data processing. To increase safety and operational effectiveness, future integration techniques will concentrate on improving detector flexibility for long-duration missions, deep space exploration, and human spaceflight [21]. Upcoming space mission's security and achievement depend heavily on the identification of radiation from space. To safeguard humans, equipment, and spacecraft from the damaging effects of radiation, accurate radiation detection and monitoring are crucial. Modern detection technologies will be essential for tracking the exposure of radiation and guaranteeing proper shielding when missions travel farther and take longer than the Earth's lowest orbit. The capacity to anticipate, identify, and react to radiation hazards within will be improved by emerging technology and integrating tactics, protecting mission objectives and public health. The viability of challenging exploration objectives and the advancement of deep spaceflight depend on efficient radiation detection [22].

8.5.2   Final Thoughts

Future developments in space radiation detection have the potential to greatly improve our capacity for safe space exploration and operation. Establishing more complicated, robust, and effective detection systems will be critical when missions go farther into space and take longer to complete. Advances in materials science, data processing, and sensor technology are expected to enhance precision and flexibility, tackling present issues and broadening the scope of radiation surveillance. To meet changing difficulties and guarantee mission success, space radiation detection requires constant innovation and attention to detail. The increasing complexity of radiation environments resulting from space travel drives the ongoing development of sophisticated detection devices and procedures. Future space missions will be reliable and efficient because of the dedication to invention and active oversight, which will open the door for ground-breaking research.

8.6   Future Perspectives

8.6.1   Vision for the Future

The long-term objectives for space radiation detection technologies include extremely sophisticated, multipurpose devices that can monitor accurately and in real-time in a variety of settings. These technologies, which make use of advanced components and are based on artificial intelligence analytics, will easily interface with spaceship and exploratory instruments. The objective is to improve safety, anticipate radiation risks, and allow successful, prolonged trips farther into space with more assurance and adaptability. Prolonged expeditions to Mars and beyond are possible space exploration situations where sophisticated radiation detection will be essential to human safety and mission success. Good detection systems might improve mission planning, facilitate deep space settlement, and reduce the health concerns associated with protracted radiation exposure. These developments will push the limits of human space exploration and habitation by enabling longer, healthier research trips.

8.6.2   Next Steps for the Field

The development of radiation sensors with improved sensibility, robustness, and analytics in real-time should be the main goal of future studies. Developing materials science to make components resistant to radiation, incorporating AI into forecasting, and validating detectors in virtual deep space settings are among the top priorities. To handle new issues in space radiation detection and accelerate innovation, cooperation between aerospace organizations and research institutions will be essential. By combining resources, knowledge, and data, global cooperation might dramatically increase the detection of space radiation. Joint research programs, shared testing spaces, and agreements for sharing information between space agencies and academic organizations are a few instances of these joint endeavours. These collaborations have the potential to improve safety and efficiency throughout the worldwide space community by promoting innovation, standardizing identification procedures, and strengthening readiness for deep space missions.

References

1. Sbrogiò G (2011) Design and construction of a microdosimetric detector for the International Space Station (ISS)
2. Ahmad MI, Ab Rahim MH, Nordin R, Mohamed F, Abu-Samah A, Abdullah NF (2021) Ionizing radiation monitoring technology at the verge of internet of things. Sensors 21(22):7629
3. Van Royen Y. The future of Moon spacewalks: next-generation radiation protective spacesuit
4. National Research Council, Division on Engineering, Physical Sciences, Aeronautics, Space Engineering Board and Committee on the Evaluation of Radiation Shielding for Space Exploration (2008) Managing space radiation risk in the new era of space exploration. National Academies Press
5. McGregor D, Shultis JK (2020) Radiation detection: concepts, methods, and devices. CRC Press
6. Cavan AE (2015) Digital holographic interferometry for radiation dosimetry
7. Benton ER, Benton EV (2001) Space radiation dosimetry in low-Earth orbit and beyond. Nucl Instrum Methods Phys Res Sect B Beam Interact Mater At 184(1–2):255–294
8. Rossi AP, Van Gasselt S (eds) (2018) Planetary geology. Springer International Publishing, Cham, Switzerland, pp 1–414
9. Henry RC (1994) Voyager investigation of the cosmic diffuse background: observations of rocket-studied locations with Voyager (No. NAS 1.26: 197532)
10. Nagihara S, Williams DR, Nakamura Y, Kiefer WS, McLaughlin SA, Taylor PT (2020) Availability of previously lost data and metadata from the Apollo Lunar Surface Experiments Package (ALSEP). Planet Space Sci 191:105039
11. Mauk BH, Haggerty DK, Jaskulek SE, Schlemm CE, Brown LE, Cooper SA, Gurnee RS, Hammock CM, Hayes JR, Ho GC, Hutcheson JC, Jacques AD, Kerem S, Kim CK, Mitchell DG, Nelson KS, Paranicas CP, Paschalidis N, Rossano E, Stokes MR (2017) The Jupiter energetic particle detector (JEDI) instrument for the Juno mission. Space Sci Rev 213:289–346
12. Hassler DM, Zeitlin C, Wimmer-Schweingruber RF, Ehresmann B, Rafkin S, Eigenbrode JL et al (2014) Mars' surface radiation environment measured with the Mars Science Laboratory's Curiosity rover. Science 343(6169):1244797

13. Seco J, Clasie B, Partridge M (2014) Review on the characteristics of radiation detectors for dosimetry and imaging. Phys Med Biol 59(20):R303
14. Guo J, Zeitlin C, Wimmer-Schweingruber RF, Hassler DM, Ehresmann B, Rafkin S, Freiherr von Forstner JL, Khaksarighiri S, Liu W, Wang Y (2021) Radiation environment for future human exploration on the surface of Mars: the current understanding based on MSL/RAD dose measurements. Astron Astrophys Rev 29:1–81
15. Hyde RA, Ishikawa MY, Nuckolls JH, Wood LL (2003) Optical power-beaming from satellite power-stations: economic imperatives and provision of high value-adding electrical power. Tech Rep
16. Hassan HS, Elkady MF (2020) Semiconductor nanomaterials for gas sensor applications. Environ Nanotechnol 3:305–355
17. National Research Council, Division on Engineering, Physical Sciences, Standing Committee on Technology Insight—Gauge, Review, and Committee on Developments in Detector Technologies (2010) Seeing photons: progress and limits of visible and infrared sensor arrays. National Academies Press
18. Post MA, Yan XT, Letier P (2021) Modularity for the future in space robotics: a review. Acta Astronaut 189:530–547
19. Kato N, Fadlullah ZM, Tang F, Mao B, Tani S, Okamura A, Liu J (2019) Optimizing space-air-ground integrated networks by artificial intelligence. IEEE Wirel Commun 26(4):140–147
20. Narici L, Berger T, Matthiä D, Reitz G (2015) Radiation measurements performed with active detectors relevant for human space exploration. Front Oncol 5:273
21. Sandhu HK, Bodda SS, Gupta A (2023) A future with machine learning: review of condition assessment of structures and mechanical systems in nuclear facilities. Energies 16(6):2628
22. Dobney W, Mols L, Mistry D, Tabury K, Baselet B, Baatout S (2023) Evaluation of deep space exploration risks and mitigations against radiation and microgravity. Front Nucl Med 3:1225034